Science and Technology in The Mahābhārata

By

Dr. Ravi Prakash Arya

AMAZON BOOKS, USA

in association with

INDIAN FOUNDATION FOR VEDIC SCIENCE
1051, Sector-1, Rohtak-124001, Haryana, India
Ph. Nos.: 09313033917; 09650183260
Emails: vedicscience@rediffmail.com
vedicscience@hotmail.com
Website : www.vedicscience.net

First Edition

Kali era: 5015 (c. 2014)
Kalpa era: 1,97,29,49,115
Brahma era: 15,50,21,97,9,49,115

ISBN No. 1482571978

Table of Contents

Introduction

The *Mahābhārata* is the turning point not only in the history of India but in the history of the whole world. The Mahābhārata war was a death stroke to the Indian scientific advancement, prosperity and glory. Before this catastrophe, Indian civilisation was in full vigour. The Mahābhārata war was India's loss and world's gain. This war deprived India of her highly advanced science and technology due to mass scale destruction of her great men in the war and mass scale migration of those skilled men survived the war to the various parts of the globe. Thus India's greatness began to decline and the entire western world received in colonists the seeds of their future scientific advancement and greatness. The *Mahābhārata* was so fraught with worldwide consequences. Mr. Pococke (the author of *India in Greece*, p. 26) has so rightly remarked on this fact. According to him,''But, perhaps, in no similar instance have events occurred fraught with consequences of such magnitude, as those flowing from the great religious war which, for a long series of years, raged throughout the length and breadth of India. That contest ended by the expulsion of vast bodies of men, many of them skilled in the arts of early civilisation and still greater numbers warriors by profession. Driven beyond the Himalayan in the north ....The mighty human tide that passed the barrier of the Punjab, rolled onward towards its destined channel in Europe and Asia, to fulfil its beneficent office in the moral fertilisation of the world.''

Thus the study of the *Mahābhārata*, the only surviving historical document, enables one to know about the scientific and cultural legacy of great India. This paper tries to unravel the scientific and technological advancement of India at the age of the *Mahābhārata*, i.e. 5000 years ago when Indian civilisation was at its pinnacle. This study would help proper portrayal of India's past scientific achievements thereby recognition of India's contribution to the world civilisation in science and technology.

Study of the *Mahābhārata* has a great relevance and significance in the present day context. It acts as a link between the past and present not only in India but also in the entire world. The *Mahābhārata* period has a great significance in the history of Bhārata. The *Mahābhārata*, in fact, is the sheet anchor in the history of our country. It was the period when the great scientific advancement of the Vedic period preserved safely and the entire Vedic knowledge was revised and re-edited by the great seers of that period. Thus, the the the *Mahābhārata* proves to be a significant tool to understand India's hoary past. The economy, social and political structure, scientific advancement and what not.

This period can also be described as the transition period in the history of Bhārata when the entire scientific knowledge and technological know-how of Vedic period were destroyed in the the Mahābhārata war. Third but not the least significance of the Mahābhārata period is that this war marked the end of the Dvāpara Yuga and the commencement of the new era of Kaliyuga. The Mahābhārata war leads to the destruction of entire scientific knowledge and technological know-how that was handed down to us traditionally. We were left only with cryptic formulae and theories related to the system of knowledge, scientific advancement and technological know-how, historical mentions of which is found only in the surviving texts. The mere historical mention of a system of knowledge, scientific advancement and technological know-how in the surviving texts would appear to the future generations devoid of this type of knowledge as merely nonsense until such times that these theories were discovered a fresh. Merely philological analysis without the knowledge of the cultural and scientific background within which the concerned texts were composed will be inadequate in any process of decipherment of these texts. This was the actual case with our ancient texts. That was the reason why the academic scholarship of 49th (eighteenth c. B.C) and 50th

(nineteenth c. B.C) century of Kali era failed in unravelling the mysteries of the ancient Sanskrit texts for want of any scientific discovery at par with those of the Mahābhārata war in the 49th and 50th centuries of Kali era. 51st century of Kali era has witnessed a lot of scientific advancement. This advancement also led to the rediscovering of such theories as were enshrined in our ancient Sanskrit texts and the various theories and their applied aspects that appeared to the scholars till 51st century of Kali era (20th century of Christ) as mere nonsense and false appear to be facts and realities. Until 50th century of Kali era (19th century of Christ) when we had no idea of nuclear weapons or no means to know the radiation level, the Mahābhārata war appeared to most of us as imaginative or magical one. However, as soon as the nuclear weapons were discovered and techniques of measuring radiation levels were explored, we have become able to say that the Mahābhārata war was a nuclear war. This fact was confirmed by Dr. S.K. Trikhā, Head of the Physics Dept. of Delhi University in his research paper read by him in a seminar organised by Rashtriya Sanskrit Sansthan from 11 October 1995 to 15 Oct. 1995 to celebrate its silver Jubilee function. Dr. Trikhā proved in his paper that nuclear weapons were used in the great war of the Mahābhārata. In support of his contention, he located some such spots around Kurukshetra where the radiation level is 2.5 times higher than the normal level even at this time when 5100 years have elapsed the occurrence of war. This level of radiation is said to be equal to the level of radiation at the spot of Hiroshima and Nagasaki in Japan where the American forces dropped the atomic bombs during the Second World War in 1944. Thus, it became a proved fact that the Mahābhārata war was a nuclear war that took place in Kurukshetra. The power of nuclear weapons can be imagined by the radiation level that even after 5000 years equals to the radiation level of Hiroshima and Nagasaki where the atom bomb was dropped only 60 years ago.

Significance of Science

During Mahabharata period (Śānti Parva, 205.3), science was given a great significance. Pointing out the great significance of science, Vyāsa says:

प्रज्ञया मानसं दुःखं हन्याच्छारीरमौषधैः ।
एतद् विज्ञानसामर्थ्यं न बालैः समतामियात् ।।

It is the significance of science that we are able to cure the psychological problems through psychotherapies and physical diseases through administration of medicines, nor we would have wept like children when faced with these diseases.

Stressing upon the importance of knowledge and its realisation or practical applicability (science), Vyāsa (Śānti Parva, 326.22) observes that without the Jñāna (knowledge) and Vijñāna (science), it is difficult to attain Mokṣa.

न बिना ज्ञानविज्ञाने मोक्षस्याधिगमो भवेत् ।

In Anuśāsana Parva of the *Mahābhārata* (8.8), Bhiṣma Piāmaha while narrating the qualities of scholars gives utmost importance to those scholars who are well versed in science.

ये चापि तेषां श्रोतारः सदा सदसि सम्मताः ।
विज्ञान गुण सम्पन्नास्तेभ्यश्च स्पृहयाम्यहम् ।।

Astronomy or Astrophysics

In the age of the Mahābhārata, we find that people were fully acquainted with our solar family. Hindus had the detailed knowledge of our planetary system. People were quite familiar with the planets like Uranus and Neptune by the names of Śveta and Śyāma. The great sage Vyāsa named them so while narrating the positions of all the planets at the beginning of the great war. The names given to the three planets show that their colours were also known to the Indians at the age of the Mahābhārata itself i.e. 5100 years ago, whereas the rest of the world has come to know about it only recently.

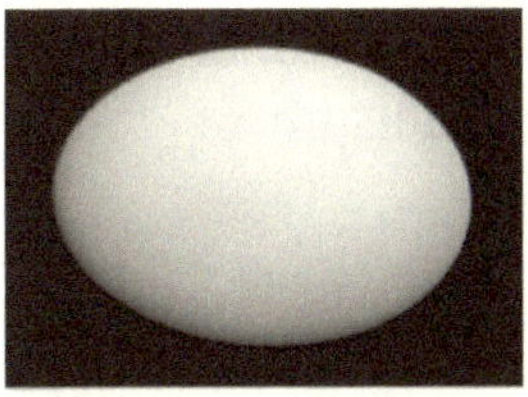

Śveta means greenish-white and the colour of Uranus as we know today is greenish-white.

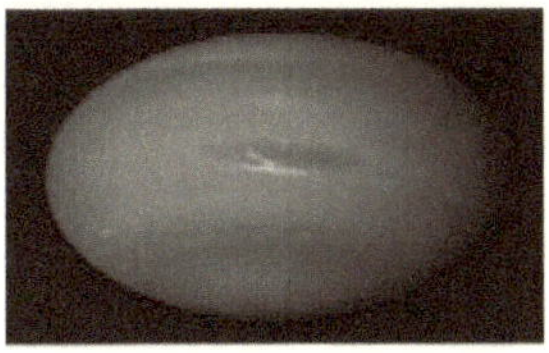

Śyāma means blue and Neptune is blue in colour according to modern science.

The rest of the world has only recently come to know, through the researches done in modern astronomy that the star Vega slipped down in the sky around 15000 years ago to occupy the position of celestial pole. However, the Hindus even at the age of the Mahābhārata were keeping the celestial event alive in their memory that was observed by them some 10,000 years ago.

Medical Sciences

Medical science was also far advance at the age of the Mahābhārata.

Test tube babies without ovum

In modern times the concept of test tube babies came into being only recently in 1979 only when Mr. Steptoe discovered the technique of producing test tube babies with the help of ovum and sperm. On the other hand test tube babies were prepared in this country since the Vedic period not only with the help of ovum and sperm only but with the help of ovum only and with sperm only. In The Rigveda (7.33.13) we find a mention of the technique of the producing a child with the help of testtube.

सत्रे ह जादविषिता नमोभिः कुम्भे रेतः सिषिचतुः समानम्
ततो ह मान उदियाय मध्यात् ततो जातमृषिमाहुर्वसिष्ठम्।।

For instance, Agasti and Vasiṣṭha were born out of an urn

called vastivara from the semen of Mitra and Varuṇa. Bhiṣma of Mahabhārata also mentions this fact to Yudhiṣṭhira.

तस्यान्तरिक्षं पृथ्वीं दिवं च सर्वं वशे तिष्ठति शाश्वतस्य।
स कुम्भे रेतः ससृजे सुराणां यत्रोत्पन्नमृषिमाहुर्वसिष्ठम्।।
महा. अनु.पर्व. 158.19

The Mahābhārata period was no exception to this achievement. Draupadi and Dhṛṣṭdyumna got birth from a utensil only from the semen of king Drupada. Droṇa was born from the semen of Bharadvāja in an urn known as Droṇa.

तत्र संसक्तमनसो भरद्वाजस्य धीमतः।
ततोऽस्य रेतश्चस्कन्द तदृषिर्द्रोण आदधे।।
ततः समभवद् द्रोणः कलशे तस्य धीमतः।
अध्यगीष्ट स वेदांश्च वेदानि च सर्वशः।। महा.आदिपर्व. 129. 37–38

Kṛpa and Kṛpī were twins born from the semen of Gautama in Śarastambh. All the above-cited test tube babies were developed from only the semen without the use of ova, the female part. Modern science has not yet produced test tube babies using sperm, male part only. Modern science is capable of developing test tube babies with the help of both male and female parts only. Such an experiment is yet to be done by the present day scientists. If we are able to achieve success in developing test tube babies without ovum like that of the Mahābhārata, even a male person will be able to have children without having married to a female partner.

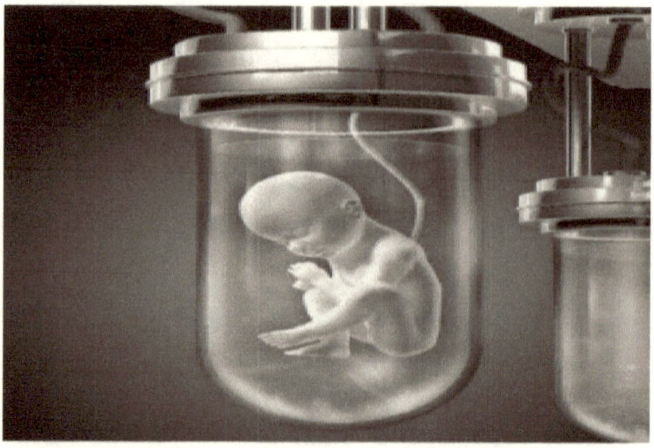

Developing embryo in vitro

During the Mahābhārata Vedic scientists were capable of developing the embryos in vitro, i.e. out of any living body of any animal. On the other hand, modern science is not capable of doing this. In the modern experiments, only fertilisation of an ovum was done in a tube, while the further growth accrued in the uterus of a living woman after implantation of the fertilised ovum in it. The baby is born out of the uterus and not out of the tube. In the *Mahābhārata*, we find that the test tube baby Droṇa was married to a test tube baby Kṛpī. Their development and behaviour were normal. They led a normal married life producing one son Aśvatthāmā. We also find that another test tube baby Kṛpa was married to a normal. This couple too led a normal married life.

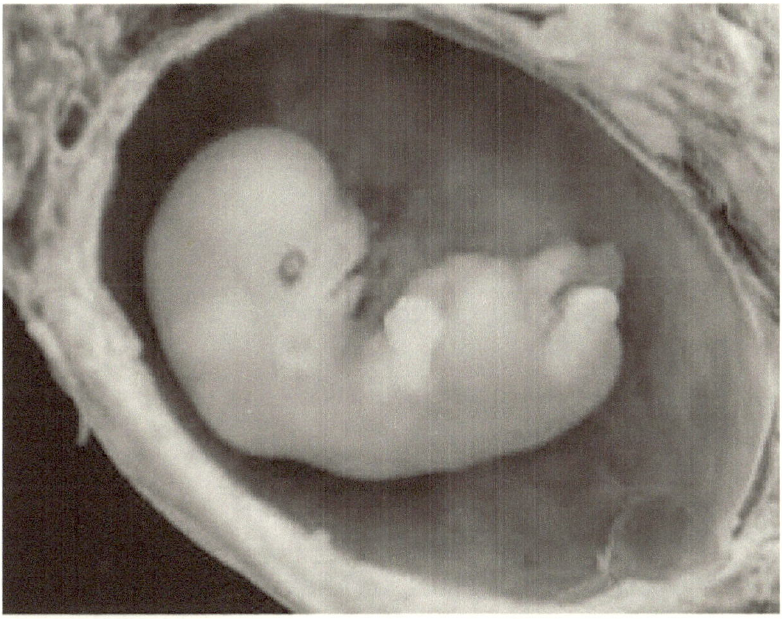

Parthenogenesis

Parthenogenesis is a form of reproduction in which an egg can develop into an embryo without being fertilized by a sperm. Parthenogenesis is derived from the Greek words for 'virgin birth'. In fact in Greek language Parthenogenesis for 'virgin birth' was borrowed from Sanskrit Pṛthājanana which also meant birth given like 'virgin Pṛthā (Kuntī) to the Pāṇḍavas. Here it may be known that several insect species including aphids, bees, and ants are known to reproduce by parthenogenesis. The Mahābhārata scientists also developed this technique. Pṛthā (Kuntī) used this technique to produce Pāṇḍavas. She stimulated her ovum with the help of solar energy and produced Karṇa. Similarly, other energies were used to produce other children. This technique may also be developed in the present age also to update the modern knowledge of medical science for the benefit of human beings who need this technique to produce chidren.

Parthenogenesis:

- **is a form of asexual reproduction in which females produce eggs that develop without fertilisation**
- **is virgin birth**

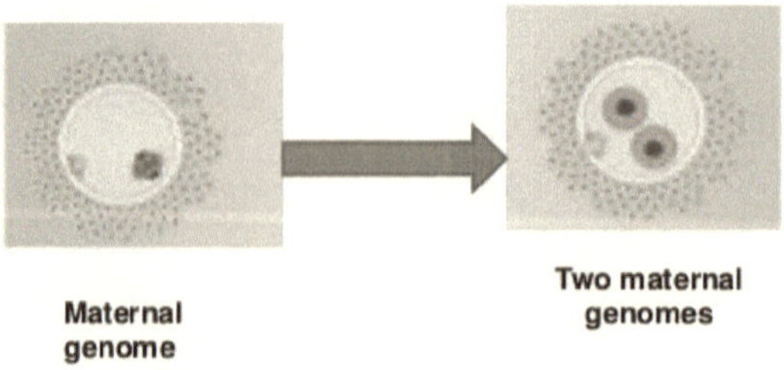

Determination of sex

The Mahābhārata scientist knew that sex is determined by the rate of metabolism and quality of food. Nevertheless, this question is still a puzzle before modern science. Insects like wasps, bees, and ants do produce males when the eggs are not fertilised by the sperms of the male. Durvāsā, a visionary sage of the *Mahābhārata* solved this puzzle during his experiments on Kuntī by suggesting her regimen be followed by her during her pregnancy.

Cloning, and embryo transfer

If you had read the Mahābhārata a hundred years back, you would have thought all those advanced medical descriptions should be "divine" and not humanly. Now in this 21st century, we can perceive them to some extent with scepticism due to the development in science. Therefore, more and more we advance in science we could perceive the advanced scientific concepts of the *Mahābhārata* better.

There is a mention of an embryo, conceived in one womb, being transferred to the womb of another woman from where it is born. The transferred embryo is Balarama and this is how he is a brother to Krishna although he was born to Rohini and not to Devaki.

Karna the son of Surya (sun god) is born with a shield that protects him from weapons. This may refer to the science of genetics where modification of DNA is done bio-technically to provide a biological shield similar to Exoskeleton found in tortoise but incredibly strong. Unlike artificial Exoskeleton used by humans, Karna's Kavacha was blended with his body.

Multiplication of fertilised Ovum

The Mahābhārata scientists knew that a fertilised ovum multiplies repeatedly to give rise to many cells. Up to a certain stage, all the new cells possess the capability to multiply from the beginning to produce full-grown offspring. The celebrated sage Vyāsa applied this knowledge while taking possession of the aborted embryo of Gāndhārī. He examined the aborted embryo and separated the normal cells from the aborted mass, which was a mole .i.e abnormal growth. He implanted those normal cells in a Ghṛta Kumbha, i.e. a pot containing a nutrient medium. Thus 100 Kauravas and a daughter Duśśalā were born from the normal cells of the aborted mass that grew in vitro, i.e. the medium outside the uterus. All these records show that how developed medical science was in the Mahābhārata. Modern scientists can follow the experiments and techniques to update their knowledge in the context of the ancient science of Hindus.

Surgery

Ancient India had great physicians and surgeons who were far more advanced in their profession. For instance, Suṣruta was known as "father of surgery" did Cataract surgery. He used a special tool called the "*jabamukhi śalākā*", a curved needle, which was used to loosen the obstructing phlegm and push it out of the field of vision. In the Mahābhārata, surgeons and surgery played a very important role in war and life. The role of surgeons was mentioned in various instances. In Bhiṣ ma parva, Sañjaya said, "Then there came unto him some surgeons well trained (in their science) and skilled in plucking out arrows, with all becoming appliances (of their profession). Beholding them, the son of Gaṅgā said unto thy son,--'Let these physicians, after proper respect being paid to them, be dismissed with presents of wealth. Brought to such a plight, what need have I now of physicians? I have won the most laudable and the highest state ordained in Kṣatrīya observances! Ye kings, lying as do on a bed of arrows, it is not proper for me to submit now to the treatment of physicians. Udyoga parva (section 152) informs us that Kunti's son, king Yudhishthira, marched, taking with him the cars and other vehicles for transport, the food-stores and fodder, the tents, carriages, and draught-cattle, the cash-chests, the machines and weapons, the surgeons and physicians, the invalids, and all the emaciated and weak soldiers, and all the attendants and camp-followers. In the same parva (Udyoga parva, section 153) it is informed that there assembled hundreds upon hundreds of skilled mechanics, in receipt of regular wages and surgeons and physicians, well versed in their own science, and furnished with every ingredient they might need. Śānti parva (section 95) informs that either a wounded opponent should be sent to his own home, or, if brought to the victor's quarters, should have his wounds attended to by skilful surgeons. When in consequence of a quarrel between righteous kings, a righteous warrior falls into distress, (his wounds should be attended to and) when cured he should be set at liberty.

ESP in Womb

The *Mahābhārata* talks about extra sensory perception in the womb. When Subhadrā was pregnant, Arjun told her the secret of entering the Cakravyūha. When he was explaining the exit procedure, Subhadrā fell asleep. Hence, Abhimanyu learned the entrance procedure while he was in Subhadra's womb and he never had the chance to learn the exit strategy.

There was a time when people criticised this concept of a child learning from the womb of a mother. Further scientific developments have made things very clear and modern science confirms its possibility. In his book "Right Brain Education in Infancy" by Dr. Makoto Shichida, world-renowned founder of over 350 Child Academies in Japan says, the right brain is active during gestation. Furthermore, the imaging right brain is the centre for extrasensory perception or ESP. He further adds cells are sensitive and can transmit subtle energy patterns to the brain; a child has an extra sensory perception in the womb. The unborn baby uses the cells of his developing body to gain information from the world around him, which is transmitted to his right brain hemisphere. Unless there is interference, the right brain hemisphere is capable of putting extrasensory impressions on the screen of the mind, which is how it got the name image brain. Young children are highly receptive when it comes to extrasensory impressions because the left-brain is not dominant in children until the age of six or seven. Dr. Shichida found that young children can be easily trained in ESP and that babies in the womb have ESP. He believes this is because the foetal right brain hemisphere is active while the left-brain hemisphere is dormant. Source: http://shichida.co.jp/

The existence of Bacteria and Viruses

The existence of bacteria and viruses are explained in the *Mahābhārata* (*Śānti parva*, Section, 15).

> Arjun said, "I do not behold the creature in this world that supports life without doing any act of injury to others. Animals live upon animals, the stronger upon the weaker. The mongoose devours mice; the cat devours the mongoose;

the dog devours the cat; the dog is again devoured by the spotted leopard. Behold all things again are devoured by the Destroyer when he comes. This mobile and the immobile universe are food for living creatures. This has been ordained by the gods. The very ascetics cannot support their lives without killing creatures. In water, on earth, and fruits, there are innumerable creatures. It is not true that one does not slaughter them. What is higher duty there than supporting one's life? Many creatures are so minute that their existence can only be inferred. With the falling of the eyelids alone, they are destroyed"

Advanced Vision

Modern science clearly states the limitation of our human eye. The eye has limited size and therefore limited light-gathering power. Ophthalmologist says that the human eye has limited frequency response and it can only see electromagnetic radiation in the visible wavelengths. Therefore, what we see is not what exists but what we will be capable of seeing. That is why lord Krishna provided Arjuna with special visionary powers (divine eyes) to enable him to have a glimpse of his " *Viśvarūpa*" (exposing his cosmic body).

Modern science enables refractive laser eye surgery, performed by ophthalmologists for correcting myopia, hyperopia etc. Laser energy was introduced in refractive surgeries discovered by Rangaswamy Srinivasan. Srinivasan working at IBM Research Lab discovered that an ultraviolet Excimer laser could etch living tissue in a precise manner with no thermal damage to the surrounding area. Here Laser energy is applied to the eye for reshaping, and the flap is replaced to serve as a type of natural bandage for quicker healing. In Arjun's case, too, lord Krishna might have used an advanced mode of laser-like energy to enable Arjun's eye to visualise extreme frequencies.

Continuous change of particles in the body

Please note, the continuous change of particles in the body was well known to the sages long before. In the modern world, it was shown that William Harvey (1578 - 1657) discovered the circulation of the blood. The *Mahābhārata* provides a clear view about the same as follows:

> "The constituent elements of the body, which serve diverse functions in the general economy, undergo change every moment in every creature. Those changes, however, are so minute to be noticed clearly. The birth of particles, and their death, in each successive condition, cannot be marked, O king, even as one cannot mark the changes in the flame of a burning lamp. When such is the state of the bodies of all creatures, - that is when that which is called the body is changing incessantly even like the rapid locomotion of a steed of good mettle- who then has come whence or not whence, or who is it or who is it not or whence does it not arise? What connection does there exist between creatures and their own bodies?"

Life Science

Life science also remained in a fully developed state until the Mahābhārata period.

Discovery of Chromosomes

Chromosomes were known to Hindus until 5000 years ago, whereas modern science could rediscover these principles only by 4991 Kali era (1890 BC). The Mahābhārata scientists termed them as *guṇavidhi.* This was exactly a scientific name because chromosomes control guṇas (characters) and vidhis (functions) of a person. There being 23 in number in case of human beings was also known by the time of the *Mahābhārata.* The references of the *Mahābhārata* and Bhāgwata prove this fact. It is also stated that these principles give rise to genetic diseases. A list of genetic diseases is cited in the *Mahābhārata,* which tallies to list made by modern science.

Fertilization process

Medical Science during the Mahabharata period was well acquainted with the fertilization process. Accordingly,

> बिन्दुन्यासादयोऽवस्थाः शुक्रशोणितसम्भवः ।
> यासामेव निपातेन कललं नाम जायते।।

कललाद् बुदबुदोत्पत्तिः पेशी च बुद्बुदात् स्मृता।
पेश्यास्त्वाभिनिर्वृत्तिर्नखरोमाणि चा तः।।
सम्पूर्ण नवमे मासि जन्तोर्जातस्य मैथिल।
जायते नामरूपत्वं स्त्री पुमान् वेति लिंगतः।।

– महाभारत, शान्तिपर्व 320.117–119

When ovum and sperm combine to form a zygote (Kalala). zygote is divided into cells forming a ball shaped structure (buda-buda) called Morula. Morula gives rise to the Foetus (pesi) formation. Fetal hair and nail start growing around 11 weeks of pregnancy and a fully-grown baby takes birth during the ninth month of pregnancy. After the birth, baby's naming ceremony is performed according to her male or female sex.

Zygote Formation

As per records of Bhāgvata 23 guṇavidhi of a male gamete unite with her 23 guṇavidhi to stimulate them to form a *Kalala* (zygote). Then the zygote divides once, ten times, three times. It means that cell divides once to give rise to two cells. This happens ten times so that 10^2 cells form.

Genetic Inheritance

Medical science during the period of the *Mahābhārata* discovered that The chromosomes contain the genes we inherit from our parents. Each cell in the body contains 23 pairs of chromosomes. One chromosome from each pair is inherited from mother and one is inherited from father.

Accordingly, a baby inherits bones, bone marrow, muscles and nerves from father. The blood, flesh and skin is inherited from mother.

ये गुणाः पुरूषस्येह ये च मातृगुणास्तथा।
अस्थि स्नायुश्च मज्जा च जानीमः पितृतोगुणाः।।
त्वग्मासं शोणितं चेति मातृजान्यपि शुश्रुम्।

– महा. भा. शान्ति पर्व 305.1–5

Cell Division

This happens thrice, at three layers, viz. endoderm, mesoderm and ectoderm. This multiplication produces 30^2

cells, which are present in a newly born baby. Vedic sages achieved such a vast and microscopic knowledge quite ago. Whereas, modern science could achieve this knowledge only in the 51st century of Kali era, i.e. 20th century of Christ.

Parachute for floating in the sky

Parachutes were in use for floating in the sky. The Mahābhārata tells us that Śalva air raided Dvārkā. Kṛśṇa cut the airplane namely 'śaubha' by Sudarśa cakra. Śaubha than fell down burning. Nevertheless, Śalva remained floating in the sky. Here one thing is also clear that Indians during ancient times had developed the technique of piercing the airplanes of the enemy with the weapon called Śudarśana disc. We have not been able to develop such a technique today.

Immunity to radiation

Examples of immunity to radiation are available in the *Mahābhārata.* Effect of radiation of a nuclear weapon Brahmāstra on the son of Abhimanyu caused his stillbirth, but Kṛṣṇa nullified the effect and the child regained his life. Immunity to radiation was also shown by Vyāsa, Nārada, etc. It was considered unbelievable earlier, but now modern science has established that such immunity is possible.

Elongation of life in space travel

During Mahābhārata period it was accepted as a proved fact that space travel with tremendous speed may cause elongation of life. The story of Kakudmi and Revati, the wife of Balaram portrays this fact. Revati and her father Kakudami returned from space after 26 Yugas, still, Revati was youthful, in puberty and Kakudamy lived to give her in marriage to Balaram. This fact has not yet been proved by modern science, though Einstein has supported this concept theoretically in the 51st century of Kali era.

The science of Aeronautics

Science of aeronautics remained developed to its advance stage until *Mahābhārata* period. According to the description available in *Droṇaparva*, Vimānas were shaped like a sphere and could move along at great speed on a mighty wind generated by mercury. It says the vimāna moved like a UFO, going up, down, backwards and forwards as the pilot desired.

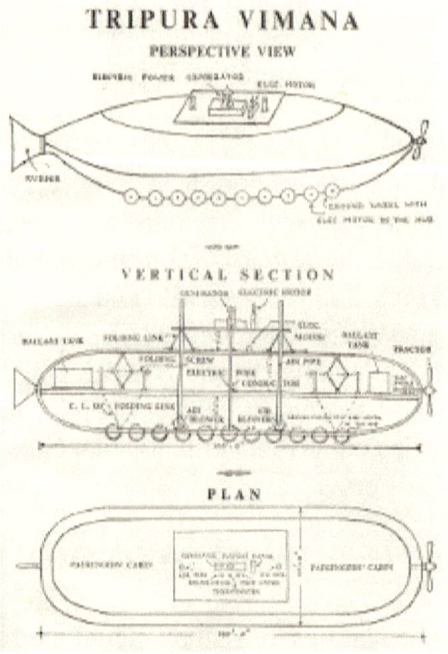

Section 43 of *Vanaparva* describes Arjuna's arrival at the city of Indra 'Amaravati'. Here not only vimānas are mentioned but even an airport, where vimānas were properly stationed and used to land and take off in a similar manner as we see today in busy airports. The description is quoted below:

> "And having beheld those celestial gardens resounding with celestial music, the strong-armed son of Pāṇḍu entered the favourite city of Indra. In addition, he beheld there celestial vimānas by thousands, capable of going everywhere at will, stationed in proper places. And he saw tens of thousands of such cars moving in every direction."

Also, section 165 of *Vanaparva* mentions about aircraft as follows:

> "And from all sides in vimānas resplendent as the sun, hosts of Gandharvas and Apsaras began to follow that repressor of foes, the lord of the celestials. In addition, ascending a vimāna yoked with engines powerful like a horse, decorated with burnished gold, and roaring like clouds, that king of the celestials, Purandara blazing in beauty came unto the Pārtha. And having arrived (at that place), he guarded by a thousand eyes of security men descended from his vimāna".

See also the description (section 164) when Arjuna was dropped by Śiva's space ship piloted by "Mātali".

> "And it came to pass that once days as those mighty charioteers were thinking of Arjuna, seeing Mahendra's vimāna, yoked with engines with the effulgence of lightning, arrive all on a sudden, they were delighted. And driven by Mātali, that blazing vimāna, suddenly illuminating the sky, looked like smokeless flaming tongues of fire, or a mighty meteor embosomed in clouds and seeing before them that vimāna driving in which the slayer of Namucī had annihilated seven phalanxes of Diti's offspring, the magnanimous Pārtha went round it. In addition, being highly pleased, they offered excellent worship unto Mātali, as unto the lord of the celestials himself. Moreover, then the son of the Kuru king duly enquired of him after the health of all the gods and Mātali also greeted them. And having instructed the Pārtha even as a father doth his sons, he ascended that incomparable vimāna, and returned to the lord of the celestials".

Here the mention of "smokeless flaming tongues of fire" could be clearly perceived as an eco-friendly jet engine's emission.

War Technology in *Mahābhārata*

A careful examination of the Mahābhārata show that

different types of offensive and defensive weapons were used in during the Mahabharata period. Their operation was fully under control of the operator. The operator was equipped with such capabilities as could withdraw the operate weapon if the need be. The missiles generally returned to their point of operation after successfully hitting the target. Māntrika operation was generally preferred instead of Yāntrika or mechanical operation. Now-a-days we have not been able to develop Māntrika science and technology as it was in ancient India. We are depending upon the mechanical operation or at the most computerized electronic operation of the weapons. Mechanically or electrically controlled weapons can be operated by any person. They may even fall into wrong hands leading to the disaster of humankind. But weapons controlled through mental waves of operators cannot be handled by unwanted persons. They can more safely be preserved.

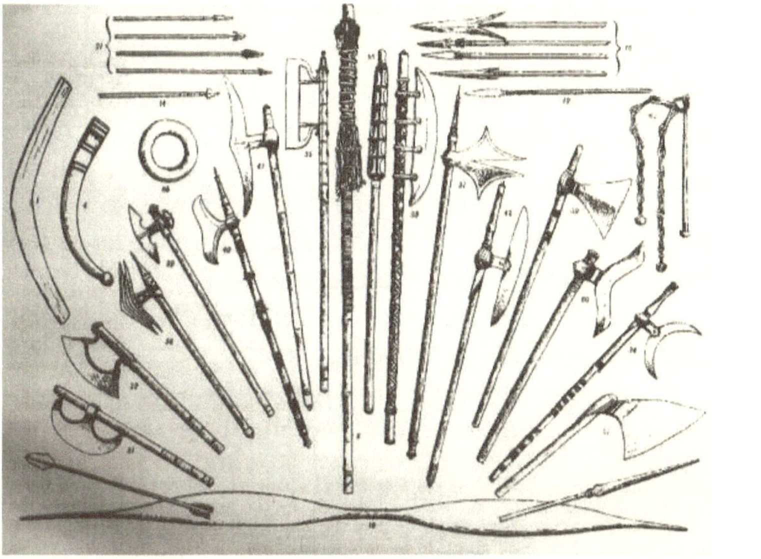

As regards the weapons used by the Indians, Professor Wilson is assured that the Indians cultivated archery most assiduously, and were masters in the use of the bow on horse-back. Their skill in archery was wonderful. "Part of the archery practice of the Hindus consisted in shooting a number of arrows at once, from four to nine at one time." Arjuna's feats in archery at the tournament before Draupadi's marriage, and again on the deathbed of Bhishma, must excite universal admiration.

Below is given a detailed break up of the equipments and weapons along with their classificaton used by armies of ancient India till the period of the Mahābhārata.

Classification of weapons

In Ancient India, weapons were divided into two types : *Astras* and *Śastras.* Astras were those weapons whose operation depended upon a *mantra* (click of a particular idea in the mind) or a machine (*yantra*) or by firing the Agnichūrṇa (gunpowder).[1] Rest of weapons were called as *Śastras.*

Again *Astras* were divided into two types. *Māntrika* and Tubular. In *Śukranīti* it has been advised that if a king does not have *māntrika* weapons, he should use tubular ones to insure his victory along with other *Śastras.*[2]

Tubular weapons have been further divided into two types according to the small and large sizes of the operating machines.[3] Small sized barrelled machines were having a barrel with an oblique (horizontal) and straight (perpendicular) hole at the origin (breech). It had the length of five vitastis (45 inches), a sharp point both at the forefront (muzzle) and at the origin, which can be used in aiming at the objective, which has fire produced by the pressure of machine, contains ball made of stones or iron and powder at the origin has good wooden handle at the top, (Butt) has an inside hole of the breadth of the middle finger, holds gunpowder in the interior and has a strong

---

1 अस्यते क्षिप्यते यत्तु मन्त्र-यन्त्राग्निभिश्च तत्।
अस्त्रं तदन्यतः शस्त्रमसिकुन्तादिकं च।। शुक्रनीति- 4.7.191

2 अस्त्रं तं द्विविधं ज्ञेयं नालिकं मान्त्रिकं तथा।
यदा तु मान्त्रिकं नास्ति नालिकं तत्र धारयेत्।।
सह शस्त्रेण नृपतिर्विजयार्थं तु सर्वदा। शुक्रनीति- 4.7.192

3 नालिकं द्विविधं ज्ञेयं बृहत्क्षुद्रविभेदतः। शुक्रनीति- 4.7.194

rod.[4] This small sized weapon was nothing but the modern gun. The striking range of the tubular weapons increases with the increase in the width of the hole of the barrel and increase in the size of ball length and width wise.[5]

Large sized tubular weapon was like that of a modern cannon. According to the description of *Śukranīti*, the large sized tubular weapon is that which has post or wedge at the origin or breech, and according to its movements, can be pointed towards the aim has a wooden frame and drawn on a carriage.[6]

According to *Śukranīti*, war fought with tubular weapons is said to be Āsura or Māyika (wizardly) war.[7] War faught with the *māntrika* weapons (operated at the click of a particular idea in mind) is said to be a divine war.

The following fivefold classification of Indian weapons is exhaustive: (1) Missiles thrown with an instrument or engine called *yantramukta*; (2) Those operated mannualy were called as *hastamukta*; (3) Weapons which may or may not be thrown, e.g. *muktāmukta*, as javelins, tridents etc. ; (4) Which are not thrown, as swords, maces, etc.; (5) Natural weapons, as fists, etc.

Broadly weapons were divided into three categories: 1. *Mantra Mukta* (operated by mantra), 2. *Yantra mukta* (operated by machine) and 3. *Hasta mukta* (operated mannually). Now we give here a detailed descriptio of weapons categorised in the above three categories.

---

4 लघुदीर्घाकारधारभेदैः शस्त्रास्त्रनामकम्।
प्रथयन्ति नवं भिन्नं व्यवहाराय तद्विदः।। शुक्रनीति- 4.7.193

5 यथा दीर्घं बृहद्गोलं दूरभेदि तथा तथा। शुक्रनीति- 4.7.198

6 मूलकीलभ्रमाल्लक्ष्यसमसन्धानभाजि यत्।
बृहन्नालिकसंज्ञं तत् काष्ठबुध्नविनिर्मितम्।। शुक्रनीति- 4.7.199

7 अस्त्रं तु द्विविधं ज्ञेयं नालिकं मान्त्रिकं तथा। शुक्रनीति - 4.7. 192
मन्त्रस्त्रैर्दैविकयुद्धं नालास्त्रैस्तथासुरम्। शुक्रनीति - 4.1053

Mantra Mukta (_Mantra_ operated weapons) or Weapons of Mass Destruction or Atomic weapons

We find that different types of offensive and defensive weapons of mass destruction were also used in ancient India. Their operation was *māntrika* and was fully under the control of the operator. The operator was equipped with such capabilities as could withdraw and operate weapon if the need be. The missiles generally returned to their point of operation after successfully hitting the target. *Māntrika* operation was generally preferred instead of Yāntrika or mechanical operation. Now-a-days we have not been able to develop *Māntrika* science and technology as it was in ancient India. We are depending upon the mechanical operation or at the most computerized electronic operation of the weapons. Mechanically or electrically controlled weapons can be operated by any person. They may even fall into wrong hands leading to the disaster of humankind. But weapons controlled through mental waves of operators cannot be handled by unwanted persons. They can more safely be preserved.

Here it may be known that *mantra* operated weapons were known as *Divyāstras* or Divine weapons. In the Rāmāyaṇa and the *Mahābhārata*, there are evidences of Nuclear weapons used in the war. In the *Mahābhārata* we find a mention-

> A swift flying powerful vimana (aircraft) hurled a single projectile (rocket) charged with the power of the nuclear device. An incandescent column of smoke and flame, as bright as ten thousand suns, rose with all its splendor.

It was an unknown weapon, an iron thunderbolt, a gigantic messenger of death, which reduced to ashes the entire race of the Vshnis and the Andhakas.

> The corpses were so burned as to be unrecognizable. Hair and nails fell out; Pottery broke without apparent cause,and the birds turned white.

> ...After a few hours all foodstuffs were infected...to escape from this fire the soldiers threw themselves in streams to

wash themselves and theirequipment."

There is also evidence of Nuclear war from Rāmāyana.

"It was a weapon so powerful that it could destroy the earth in an instant. A great soaring sound in smoke and flames And on its sits death..."

The *Mahābhārata* quotes furher -

"Dense arrows of flame, like a great shower, issued forth upon creation, encompassing the enemy... A thick gloom swiftly settled upon the Pandava hosts. All points of the compass were lost in darkness. Fierce wind began to blow upward, showering dust and gravel.

Birds croaked madly... the very elements seemed disturbed. The earth shook, scorched by the terrible violent heat of this weapon.

Elephants burst into flame and ran to and fro in a frenzy... over a vast area, other animals crumpled to the ground and died. From all points of the compass the arrows of flame rained continuously and fiercely.

In the *Mahābhārata*, Book 8: *Karna Parva*, we find-

"It was an unknown weapon, an iron thunderbolt, a gigantic messenger of death, which reduced to ashes the entire race of the Vrishnis and the Andhakas."

Section 1 of Mausala Parva:

"When then next day came, Samva actually brought forth an iron bolt through which all the individuals in the race of the vrishnis and the andhakas became consumed into ashes. Indeed, for the destruction of the Vrishnis and the Andhakas, Samva brougt forth, through that curse, a fierce iron bolt that looked like a Gigantic messenger of death. The fact was duly reported to the king. In distress of mind, the king (Ugrasena) caused that iron bolt to be reduced to a fine powder.Men were employed to cast the powder into the sea."

The term "10000 suns" and "Super-abundant" perfectly fit into today's description of Nuclear Energy. In fact what is depicted as an Iron bolt could actually have referred to

Radioactive Iron (Iron-59 is known to be radioactive). Perhaps the usage of Iron-59 as a nuclear weapon was known to Indians in ancient times. The last sentence "to cast the powder into sea", is in good terms with the fact that water is a very effective shield for nuclear radiations. Perhaps that was the intention in casting the iron powder into the sea.

The section 2 of the Mausala parva also has-

> "The streets swarmed with rats and mice. Earthen pots showed cracks or broke from no apparent cause. At night, the rats and mice ate away thehair and nails of slumbering men" and "Fires, when ignited, cast their flames towards the left. Sometimes they threw out flames whose splendour was blue and red. The Sun, whether when rising or setting over the city, seemed to be surrounded by headless trunks of human form. In cook rooms, upon food that was clean and well-boiled, were seen, when it was served out for eating, innumerable worms of diverse kinds."

The above passage too reminds us of the effects of a nuclear explosion, as foodstuffs get infected in the aftermath of a nuclear explosion. In section 3, there is also a mention of the vaishnis taking shelter close to the sea coast. This might have as well meant that the vaisnis took shelter in the sea. As mentioned earlier, water is a natural shield to Nuclear radiations.

These above facts make one really wonder if the *Mahābhārata* was indeed a nuclear war. In aid to the above mentioned facts, India, from times immemorial knew the existence of atoms and the atomic energy.

Divine Weapons

Following is the brief description of the Divine weapons used in ancient India-

Pāśupatāstra : It was extremely powerful missile that was capable of destroying the three worlds i.e. the entire earth including equatorial and polar regions. It was irresistible and could resist all missiles simultaneously. It was the discovery of Śiva. Arjuna received its technical know-how from Śiva himself.

Brahmāstra : It was a discovery of Brahmā. It was a missile that was specially designed to carry the nuclear war-head. *Brahmāstra* is repeatedly mentioned in Sanskrit works. For its use see Śri Bhāgvat describing the fight between the son of Droṇa and Arjuna with the *Brahmāstra.* The Rev. KCM. Bannerjee in his work, " *The Encyclopaedia Bengalensis,*" says that the *Brahmāstra* was probably a piece of musketry not unlike the modern matchlocks." Madame Blavatsky, in her *Isis Unveiled,* also shows that "fire arms were used by the Indians in ancient times."[8] *Brahmāstra* in fact was an atomic weapon invented by Brahmā." Now when the nuclear weapons have been discovered and techniques of measuring radiation levels were explored, we have become able to say that nuclear wars are also the discovery of Vedic period. This fact was confirmed by Dr. S.K. Trikhā, Head of the Physics Dept. of Delhi University in his research paper read by him in a seminar organised by Rashtriya Sanskrit Sansthan from 11th October, 1995 to 15th Oct. 1995 to celebrate its silver Jublee function. Dr. Trikhā proved in his paper that nuclear weapons were used in the great war of the Mahābhārata. In support of his contention, he located some such spots around Kurukshetra where the radiation level is 2.5 times higher than the normal level even at this time when 5100 years have elapsed the occurrence of war. This level of radiation is said to be equal to the level of radiation at the spot of Hiroshima and Nagasaki in Japan where the American forces dropped the atomic bomb during the second world war in 1944. Thus it became a proved fact that the Mahābhārata war was a nuclear war that took place in Kurukshetra. The power of nuclear weapons can be imagined by the radiation level which even after 5000 years equals to the radiation level of Hiroshima and Nagasaki of Japan 60 years ago when atom bomb was dropped there by Americans.

Āgneyāstra : It was developed by Bṛhaspati who first

---

[8] *Encyclo. Bengal,* Vol. III, p. 21

taught its operation to Bhardvāja, who passed it over to Agnivesh. Agnivesh handed it over to Droṇa, who ultimately taught it to Arjuna. It was extremely dreadful and frightful (*Mahābhārata* - 5.140.6). It espoused fire all around. It was counteracted with *Vāruṇāstra.*

***Vāruṇāstra* :** It was used to shower rains upon the enemy.

***Vāyvāstra* :** Dust storms were created by the use of this weapon.

Parjnyāstra : It was used to make the weather cloudy.

Antardhānāstra : Its use could make one invisible instantaneously.

Nāgāstra* or *Nāgapāśa : This weapon was developed by Varuṇa. Arjuna learnt its use from Varuṇa only. It was used to punish the sinners. It was combated with Garuḍāstra.

***Mahendrāstra* :** It was used by Arjuna upon Karṇa, who combated it with Bhārgavāstra.

***Akṣisammārjanāstra* :** The use of this missile caused the enemy die or made him insane or sometimes made him sleep or on certain occasions made him laugh incessantly. (*Mbh.* 5.94.38-40)

Ayojāla : This divine weapon when used 'created a net of iron'.

Atharvanāstra : This missile was popularly known as killer of enemies. It was, in fact, a combat weapon for Āgneyāstra.

Asurāstra : It was a prominent missile generally used by Asuras. Ghaṭotkaca (son of Bhima) knew its operation perfectly.

Isikāstra : It was considered as the most powerful missile (*paramāstra*), unparalled producing huge flames and surrounded by a circle. Its effect could be pacified only by Brahmāstra. When Aśvatthāmā used it to destroy Pāṇḍvas, Arjuna had to employ Brahmāstra.

Gandharvāstra : It was a missile developed by Gandharvas. Arjuna got it from Tumburu and other Gandharvas.

Nārāyaṇāstra : It was a missile developed by Nārāyaṇa. Droṇācārya learnt its knowhow from Nārāyaṇa himself. Only Droṇa and Aśvatthāmā knew its operation. Surprisingly enough, even Krishan and Arjuna did not know its operation. It never returned without hitting the target. Owing to its invincibility and irresistibility, it could be employed only once. Only *Pāsupatāstra* was superior to this.

Pramohanāstra : It was also a *māntrika* missile. Its use rendered the enemy unconscious. In the period of *Rāmāyaṇa* also this missile was in currency.

Prajñāstra : It was a combat weapon for *Pramohanāstra.* It was used to bring to consciousness those rendered unconscious by the use of Pramohanāstra.

Other *Divyāstras* : Following were also enumerated as *divyāstras. Sthunākarṇa*, *Indrajāla* (witchcraft), *Saura* and *Saumya.*

Combat Weapons : During the Vedic period military science was so advance that divine combat weapons were also developed to neutralize the effect of certain Divine weapons, e.g.

Nāgāstra was counteracted with the help of *Garudāstra.*

Bhargvāstra was used to counteract the *Mahendrāstra.*

Aharavaṇāstra could be counteracted with *Āgneyāstra.*

Īṣikāstra was combated with the help of *Brahmāstra.*

Effect of ***Pramohanāstra*** was neutralized with the help of *Prajñāstra.*

Āganeyāstra was combated with the help of *Varuṇāstra.*

Māntrika Technology

Right from the Vedic period till the *Mahābhārata.*, all the three types of science and technology were in use. We come across references of war vehicles including airplanes as operated by mantra (mental power of the user) in addition to manually operated war vehicles.

On the basis of forgoing discussion, it can unhesitatingly be maintained that during the time of Vedas, Nuclear science and technology was far advance than it is today. Military science also witnessed the culmination of its advancement. This is the reason why a great annihilation of mankind and its knowledge took place. Apart from the above, we find other sciences also fully developed at the time of the Mahābhārata.

Missiles : Missiles were weapon that were operated both by *mantra* and machine. Mr. H. H. Elliot, Foreign Secretary to the Government of India (1845), after discussing the question of the use of fire-arms in ancient India, says "On the whole, then, we may conclude that fire-arms of some kind were used in early stages of Indian history, that the missiles were explosive, and that the time and mode of ignition was dependent on the pleasure of operator; that projectiles were used which were made to adhere to gates and buildings, and machines setting fire to them from a considerable distance; that it is probable that saltpeter, the principal ingredient of gunpowder, and the cause of its detonation, entered into the composition, because the earth of Gangetic India is richly impregnated with it in a natural state of preparation, and it may be extracted from it by lixiviation and crystallization without the aid of fire; and that sulphur may have been mixed with it, as it is in abundant in the north west of India."[9]

Circular Missiles : One of the seven types of wars mentioned in *Vāsiṣṭha Dhanurveda* (9) was fought with weapons of the class type of circular missiles[10]. First mention

---

9 *Bibliographical Index to the Historians of M. India*, Vol. J. p. 373.

10 धनुष्चक्रन्तु कुन्तश्च खड्गश्छुरिकां गदाम्।

सप्तमं बाहुयुद्ध स्यादेवं युद्धानि सप्तधा।। 8।।

of circular missile is attested in the *Ṛgveda*[11]. In Rāmāyaṇa also we come across the mention of *Cakra* (circular missile) several times)[12]. Śri Krishna's *Sudarśana Cakra* is quite a famous weapon in the *Mahābhārata.* When the king Sālva started bombardment upon the city of Dwarika from fighter plane called *Saubha Vimāna,* Śri Krishna used *Sudarśana* missile to destroy his plane. After hitting the target successfully, *Sudarśana* named circular missile returned it to the place of origin[13]. This missile was discovered at the time of the firing the *Khāṇḍava* named forest. *Purāṇas* have extolled the significance of this missile[14]. As per the figurative description of the *Purāṇas,* Viśvakarmā is said to be the inventor of Viṣṇu's *Sudarśana* missile, Rudra's Triśūla (Trident), Kuber's *Puṣpaka* aeroplane and Kārtikeya's *Śakti.* All these weapons were divine weapons or *mantra mukta* weapons. They could be operated at the click of a *mantra* in the mind of the operator. These all weapons were made of energy of sun just like modern day laser beams or laser guided missiles.

According to *Padmapurāṇa* (*Uttara Khaṇḍa,* Ch. 145), this missile was very fierce, emitted fierceful rays all around and used to fly at the speed of light and was able to destroy all sorts of nuclear weapons of mass destruction[15].

Harivañśa Purāṇa (Bhavi. 55.21-22) describes it among the class of Āgneyāstras which was sharp edged weapon and emited rays like that of sun. It was able to destroy and burn down all types of soldiers : cavalry, infantry and armed with

---

11 anāyudhāso asurā adevāñścakreṇa tā apa vapa ṛjiṣīn*a* (RV. 8.96.9)

12 *khadgaiśca cakrairgadābhiśca.* Rām. Yudhakāṇḍa 53.9. See also Yuddha. 109.17, 86.21, 96.26)

13 *Mahbhārata, Vanaparva,* 21.2

14 *Vāyu Purāṇa, UttarakhaṭÇa,* Ch. 145

15 अथ विष्णुमुखा देवाः स्वतेजांसि ददुस्तथा। तेनाकरोन्महादेवः ....
चक्रं सुदर्शन नाम ज्वालामालातिभीषणम्

divine weapons.

In *Nītiprakāśikā*, although generally a *Cakra* (Disc shaped missile) has been enumerated in the category of *Mukta* weapons (which were operated either manually, or by machine and mantra), but certain types of *Cakras* like *Daṇḍacakra*, *Dharmacakra* were counted in the category of *Muktāmukta* weapons, meaning weapons that could be recalled by the operator to their place of origin after hitting or without hitting the target. *Muktāmukta* weapons were under the full control of the operator. *Sudarśana cakra* was, in fact, a *Paramāstra* which could not be combated by any other combat weapon.

Auśanasa Dhanurveda has divided circular missiles into uttama (best), *madhyam* (middle) and *adhama* (third) category types in view of different number of spokes, sizes and weight.

***Uttama cakra* (best type of disc missile):** First class *cakras* should have 8 spokes. 30 *pala* (34.92 kg.)[16] is considered to be the best weight of *cakra* used by adult warriors, whereas in case of young warriors the weight of a *cakra* should not go beyond 12 *palas* (13.96 kg.).

Similarly the measurement (diameter) of a first class *cakra* should be 16 *aṅgulas* [17] (12 inches) in case of adults and 8 *aṅgulas* in case of young ones. Rim of the first class *cakra* should be around 3 *aṅgulas*, round, ornamented and heavy duty rims are considered to be the best for disc missiles.

According to *Śukranīti*, the perimeter of a disc missile should be six hands (9 feet) with its diameter of 2.8 feet.

Divine wars

In addition to the above mantra operated weaponry in war, a more astonishing thing about the wars in ancient India was

---

16 **Note :** In the measurement of heavy weights 1 *pala* was taken to be 100 *tulā* and 1 *tulā* was equal to 11.640 grams).

17 Twelve *aṅgulas* is equal to one *vitasti* or a span and 24 *aṅgulas* is equal to one *hasta* or cubit. One vitasti is equal to 9 inches.

that they were also fought in a mysterious and wizzardly way. Most often than not divine missiles were used to materilise such wars. This divine weaponry was called as *Astra Vidyā*, the most important and scientific part of the art of war in ancient India which is not known to the soldiers of our age till recently. It consisted in annihilating the hostile army by envolving and suffocating it in different layers and masses of atmospheric air, charged and impregnated with different substances. The army would find itself plunged in a fiery, electric and watery element, in total thick darkness, or surrounded by a poisonous, smoky, pestilential atmosphere, full sometimes of savage and terror-striking animal forms (snakes and tigers, etc.) and rightful noises. Thus they used to destroy their enemies.[18] The party thus assailed counteracted those effects by arts and means known to them, and in their turn assaulted the enemy by means of some other secrets of the *Astra Vidyā*. Col. Olcott 'also says, "*Astra Vidyā*, a science of which our, modern professors have not even an inkling, enabled its proficient use to completely destroy an invading army, by enveloping it in an atmosphere of poisonous gases, filled with awe striking shadowy shapes and with awful sounds." This fact is proved by innumerable instances in which it was practiced. *Rāmāyaṇa* mentions it. Jalandhar had recourse to it when he was attacked by his father, Mahādeva (Śiva), as related in the *Kārtika Mahātmya.*

Yantra Mukta (Machine operated weapons)

Apart from the above Divyāstras (*mantra*-operated weapons), many other machine operated defensive weapons were in use in ancient period, e.g.

Śataghnī : It was a small rocket capable of killing 100 or 1000 persons at a time. It was operated from the walls or the gates of the forts.

In Vedas we do not find the direct mention of *Śataghnī*, but

---

[18] *Theosophist*, March 1881, p. 124

there are references of the words like *Sahasraghnī*[19] and *Śatavadha* that gives the meaning of *śataghnī*. *A V.* (1.16.44)[20] directs the killer of cow to be shot with the help of bullet made of lead. Similarly gunshot or bullet has also been referred to in *AV.*[21] (5.18.8). In *Ṛgveda*[22] (8.69.12), *Śataghnī* has been quoted as *Sūrmī*. In *Taittirīya Saṁhita*[23] (1.5.7) and Sāyaṇa's *Bhāṣya*[24] *Sūrmī* has been used in the form of cannon or gun like weapon. *VD* (67) specifically points out the existence of *Śataghnī* and *Rañjaka* (gun-powder). Accordingly, for the protection of throne of the king cannons must be installed on the fort and lot of gunpowder should also be stored. Cannon in ancient India was known by the name of *Śataghnī*.

In *Rāmāyaṇa*[25] (5.11) also we find the mention of śataghnī. In *Bālakāṇḍa Ayodhyā* has been described as a city consisting of high towers and flags equipped with hundreds of *Śataghnīs*, which shows that cannons or machines of some kind or other were used in those days to fortify and protect citadels.

The *Rāmāyaṇa*, while describing the fortifications, says "As a woman is richly decorated with ornaments, so are the towers with big destructive machines."[26] This shows that cannons or big instruments of war like cannons, which discharged destructive missiles at a great distance, were in use at that time.

---

19 धनुबिभर्षि हरितं हिरण्ययं सहस्रघ्नि शतवधम्। अथर्व0 11.2.12

20 तं त्वा सीसेन विध्यामो यथानोऽसौ अवीरहा।

21 जिह्वा ज्या भवति कुडमलं वाङ् नालिका दन्तास्तपसाभिदग्धाः।

22 अनुक्षरन्ति काकुदं सूर्म्यं सुषिरामिव।

23 एषा वै सूर्मी कर्णिकावती। एतया ह स्म देवा असुरान् शततर्हास्तृहन्ति।
य एतया समिधमादधाति वज्रमेषैतच्छतघ्नीं यजमानो भ्रातृव्याय प्रहरति।

24 ज्वलन्ति लोहमयी स्थूणा सूर्मी। सा च कर्णिकावती छिद्रवती।
अतएव ज्वलन्तीत्यर्थः।
एकेन प्रहारेण शतसंख्याकान् मारयन्तः शूराः शततर्हाः।

25 उच्चाट्टालध्वजवतीं शतघ्नीशतसंकुलाम्।

26 *Rāmāyaṇa*, *Sundara Kāṇda*, Third Chapter, 18th verse.

In descriptions of fortresses and battles, *Śataghnis* are often mentioned. *Śataghni* literally means "that which kills hundreds at once." In Sanskrit dictionaries, *Śataghni* is defined as a machine which shoots out piece of iron and other things to kill numbers of men. Its other name in *Bṛścī Kālī* बृश्चीकाली[27]

Śataghnis and similar other machines are mentioned in the following *ślokas* of the *Rāmāyaṇa*:

Canto	3	--- --- ---	*Ślokas* 12, 13, 16 and 17
"	4	--- --- ---	23
"	21	--- --- ---	last *śloka*
"	39	--- --- ---	36
"	60	--- --- ---	54
"	61	--- --- ---	32
"	76	--- --- ---	32
"	76	--- --- ---	68
"	86	--- --- ---	22

Rāmāyaṇa says that the *Śataghnī* was made of iron. In the *Sundara Kāṇḍa,* it is compared in size with big broken trees or their huge offshoots, and in appearance said to 'resemble trunks of trees." "They were not only mounted on forts but were carried to the battle-fields, and they made a noise like thunder." What else could they, therefore, be but cannons ?

Besides the *Rāmāyaṇa*, the *Purāṇas* make frequent mention of *Śataghni* being placed on forts and used in times of emergency. The name used in *Matsya Purāṇa* is *Sahasraghani* (शत) and सहस्त्र mean hundreds and thousands or innumerable)[28]

---

27 See Raja Sir RÈdhÈ Kant Deva's *fabdakalpadruma.*

28 Śataghnī differed widely from Matvāla in that the Matvāla were roiled down from maintains, while Śataghnī was an instrument from which stones and iron balls were discharged. Jamera was another machine that did fatal injury to the enemy by means of stones. See accounts of

guns and cannons are mentioned as existing in Laṅkā, under Rāvaṇa. They were called *Nhulat Yantras.*

Commenting on the passage in the Code of Gentoo (Hindu) Laws that "the magistrate shall not make war with any deceitful machine or with poisoned weapons, or with cannons and guns, or any kind of fire-arms," Halhed says, "The reader will probably from hence renew the suspicion which has long been deemed absurd, that Alexander the Great did absolutely meet with some weapons of that kind in India, as a passage in Quintus Curtius seems to ascertain. Gunpowder has been known in China, as well as India, far beyond all periods of investigation. The word fire-arms is literally the Sanskrit *Āgneyāstara*, a weapon of fire; they describe the first species of it have been a kind of dart or arrow tip with fire, and discharged upon the enemy from a bamboo. Among several extraordinary properties of this weapon, one was, that after it had taken its flight, it divided into several separate streams of flame, each of which took effect, and which, when once kindled, could not be extinguished: but this kind of Āgneyāstra is now lost."[29] He adds "A cannon is called '*Śataghnī*, or the weapon that kills one hundred men at once,' and, that the *Purāṇas* ascribe the invention of these destructive engines to Viśvakarmā, the Vulcan of the Hindus."

In the Mahābhārata also we find the mention of a war weapon like *Śataghnī.* In the *Mahābhārata Śataghnī* has been described as *aśmaguḍā* (pelting stone balls) or *ayoguḍā* (pelting iron balls).[30] They have also been described carried by two

---

battles with Mohamed Kasim.

Halhed's *Code of Gentoo Laws, Introduction*, p. 52 See also *Amarakoṣa* and *Śabda Kalpadruma*, Vol. I, p. 16

30 (क) परिगृह्य शतघ्नीश्च सचक्रा सगुडोपलाः। वनपर्व. 284.31

(ख) शूलाभुशुण्ड्योश्मगुडाः शतघ्न्यः। द्रोणपर्व. 179.37

(ग) प्रादुरासन् महाराज कार्ष्णायसमया गुडाः।

चतुश्चक्रा द्विचक्राश्च शतघ्न्यो बहुला गदा। द्रोणपर्व. 199.18–19

wheels and some times on four wheels[31].

But the western scholars thinks that the first time cannon in India was used by Babar. The time of Babar is 1526 AD. As such according to western scholars cannon was not found in India before 1526 AD.

***Arrows* :** Arrows were machine (bow)-operated weapons. There were various types of arrows like *nārāca*, *nālīka*, *varāhakarṇa*, *kṣura*, *añjlika*, *ardhacandra*, and *vaitastika*.

Vāsiṣtha Dhanurveda (45-48), gives the characteristics of good arrow. Accordingly an arrow should neither be too thick nor too thin. It should also be either made out of unripe wood material product of vile land. The arrow should have strong wielded joints. It should be made of ripe wood and should have shining. It should not have weak joints and any cracks. The time of preparation of wood arrows has also been narrated. The appropriate time for preparation of reed arrows is autumn.

VD. (45-47) has prescribed the quality of reed for arrows. The reed which has a round and hard stem and which has been grown in favourable (sunny and fertile) place can be used for making arrows. The measurement of arrow should be two hands excluding the measurement of a fist, i.e. 30 inches. (One hand is 18 inches. If the measurement of fist is excluded it will be around 15 inches). The thickness of an arrow has been prescribed equal to that of the smallest finger of the hands. The tail end of the arrow should be shaped like the feathers of a crow, swan, brown hawk, crane, peacock, vulture, osprey. SD (51) also adds the shapes of cātaka (an extinct species of a cuckoo which lived upon the drops of rain), heron, vulture and cock or feathers made of small bells. The gap between two feathers should be six aṅgulas or five inches. But the gap between two feathers in case of Śāraṅga bow has been recommended to be 10 aṅgulas or 8 inches. These feathers should be joined strongly or tied with sinew at the rate of four

---

[31] प्रादुरासन् महाराज कार्ष्णायसमया गुडाः।
चतुश्चक्रा द्विचक्राश्च शतघ्न्यो बहुला गदा। द्रोणपर्व. 199.18–19

features per reed arrow.

The Types of arrows

VD (53-54) and SD (46-47) have divided arrows in three categories : female, male, and impotent. The one heavier towards point is female, the one heavier towards the end is male and the one equal throughout is termed as impotent. The impotent type is helpful in practice of archery, the female type is a fast runner and the male type is able to pierce an object placed at along distance.

The Manufacturing, Types and Functions of Arrow heads :

According to SD (45), the arrow head should be made purified iron, should be sharp and pointed. They should be painted with hardening substance according to their feathers.

Various types of arrow-heads have been described in VD (53,54) and SD (46,47). The shape of arrowheads differ from country to country. The arrow heads can be of awl type, razor blade type, like that of a cow's tail, crescent moon shaped, needle pointed, spear headed, teeth of calf shaped, two pronged type, ear-shaped, the beak of crow and many other types.

The various types of arrow heads have various types of functions. For instance, the *ārāmukha* or an arrow head like an awl can cut through the skin, arrowhead like razor blade (*kṣ urapra*) is used to combat the arrows of the enemy or to cut up the hand of the enemy. Gopuccha (an arrow like cow's tail) is good for hitting a target in general. The arrow shaped like crescent moon is used to sever enemy's head, neck and bow. The needle pointed arrows were used to pierce the armour of an enemy. The spear headed arrow were used to pierce the chest of enemy, while two pronged arrows were used to counteract the arrows of enemy. Calf's teeth shaped arrows were used to cut the bowstrings of enemy. *Kārṇika* (flower petal shaped) arrows were used to counter the metallic arrows of enemy. Crow's beak shaped arrows were used to pierce any pierceable object (VD 55-57; SD 48-50).

In addition to the above arrow heads, a special *gopuccha* type (shaped like cow's tail) arrow head has been prescribed in VD (58) and SD (48). According to VD, this arrow head is made of sapless wood and has got a metallic nail of three *aṅgula* length fixed at its tip. On the other hand SD describes it to be made of a pure bell metal. According to another view of VD (59), a *gopuccha* can be made by replacing the head by an iron nail. This Gopuccha arrow head was used for practices in aiming targets and archery.

The arrows made completely of metals were called as *nārāca*. Some *nārācas* had five broad wings and they were able to make anybody successful in his target (VD 65; SD 60). Gun-shots or bullets were known as small arrows. They were fired by a machine fitted with a barrel (just like modern day gun). Those small arrows or bullets were useful in shooting a target placed at very high or distant places or in war taking place in a fort (VD 66; SD 61). Thus in ancient times, the guns were also used in fighting.

The Methods of Tempering Arrow-heads

In ancient times various methods were applied for tempering of arrow-heads to enhance their devastating power. The arrowheads thus tempered were able to pierce even the unbreakable armours just like a leaf of a tree (VD 62; SD 56). Long pepper (*pippalī*), *saindhava* (rock-salt), *kuṣṭha* (medical plant named Costus Speciosus or arabicus used as remedy for the disease called *takman*, in Hindi this plant is called as *kūṭ ha*) should be ground and pounded by mixing urine of a cow to prepare a paste. That paste should be smeared over the arrow-head or a weapon and then it should be heated on fire till it becomes blue like peacock's neck. When it absorbs the entire paste, it must be quenched in water. Such weapons will become the best weapons (VD 63-54; SD 57). According to SD (58) and *Jāmadgneya Dhanurveda* (JD. p.117), the paste prepared out of five types of salt soaked in honey and mustard oil is also best for tempering. It is also indicated in VD 60-61) that when the colour of white reed plant turns yellow after receiving rain water on *Svāti Nakṣatra* day, its root becomes poisonous. This root if anointed on the arrow head (tip of weapon) act as fatal. The best way to recognise the plant is that it trembles always even in the absence of wind.

Bows : Bows were missile- operating machines. Some famous bows were Gaṇḍīva of Arjuna, Pināka of Śiva and Vijaya of Karṇa.

In *Dhanurveda*, bows have been classified differently according to their sizes, measurements, and joints.

The first class of bows is called *Divya* or Divine bows. Their measurement was prescribed five hands and half, i.e. 99 inches. The bow with the same measurement was invented by Śaṅkara in ancient times. Paraśurāma also used the bow with the same measurement. After Paraśurāma, it was used by Ācārya Droṇa. From Droṇa, its use was made by Arjuna and after Arjuna, Sātyaki started using it. Thus in *Satyayuga* Mahādeva invented it, in *Tretā,* Śri Rāma used it and in *Dvāpara,* Droṇa became its operator. Second type of classification of bows was done as *Mānuṣa* bows or human bows. A bow measuring four hands, i.e 72 inches is called *Mānuṣa* bow. The length of one hand is equal to the length of 24 fingers or 18 inches. A good quality bow either have three, five, seven and nine joints. A bow with nine joints was called as '*Kodaṇḍa*' which was considered squarely an auspicious one. A bow with four, six or eight joints was considered as inauspicious one (*Sadāśiva Dhanurveda* : verse, 21-27).

Among the categories of *Divya* or Divine bows, one was called as Śārṅga bow. It was used by Viṣṇu and considered to be the best one. This bow measured seven vitasti i.e. 63 inches. (Note : One *vitasti* is a distance between the long thumb and a little finger or between the wrist and the tip of the finger and said to be equal to 12 Aṅgulas or about 9 inches). It was made by Viśvakarmā. This bow could not be handled by anybody either in north (Note : the northern hemisphere was called as *Svarga loka* and southern hemisphere was called as *Pātāla* or *asura loka.* Equatorial regions were known as *Mānuṣ a loka*). It was handled by Viṣṇu only. In addition to *Divya* category, *Śārṅga* bow was also available in *Mānuṣa* or human category. It was said that it was developed with great efforts over many years. It also served all purposes. According to *Sadāśiva Dhanurveda* (45) its measure was seven vistasti or 63 inches, but according to *Vāsiṣṭha Dhanurveda* (46) it measured six *vistati* and half i.e 59 inches. This bow could be easily used by all category of soldiers like soldiers of cavalry, infantry, soldiers on elephant's back and charioteers. The *Mānuṣa* category of *Śārṅga* bow was made of bamboo. (SD 34-37; VD 35).

According to *Vāsiṣṭha Dhanurveda* (48) three types of material was used in manufacturing bows in ancient times. It was metals, horns and wood.

Among metals gold, silver, copper and steel were used to manufacture bows. Among horns, horns of buffalo, Śarabha (Śarabha was an octopad antelope that possess big horns and almost looks like a camel found in Kashmir. It was considered more powerful even than the elephant and lion. Now this species of antelope has become extinct. This shows that the long tradition of *Dhanurveda* in India. The animals mentioned in these books have become extinct) and Rohit (a stag) were used to develop bows. The wood useful in manufacturing bows were described as sandal, cane, *śāla* (Vatica Robusta, a valuable timber tree), *sādhāvana* (a kind of hedyserum), *kakubha* (Pentaptera Arjuna), *śālmali* (silk cotton plant), segaun (teak wood), *vanśa* (bamboo) and *Añjana* tree . (VD. 38).

The characteristics of a bow string : The strings were be made of three round threads which were free from any joints, pure, fine, very soft and polished so that these threads could with stand attack in a war.

For want of silk thread, string can be made with intestines of a deer or with the intestines of a she-buffalo or a cow.

Fine strings are to be made with skin of a goat or Gokarṇa (another variety of animal). The hair on the skin should be removed thoroughly.

Sometimes strings are prepared with the bark (outer skin) of mature bamboos (plants) and those strings are tied with silken threads for making strings that withstand stand adverse situations in war.

At the advent of the month of *Bhādra* (September) the bark of the *Arka* tree becomes commendable for making strings and hence hard and sacred strings were made with it.

The threads which were obtained from the barks of the *Arka* tree are eighteen cubits in length and these should be made in triple-ply to make a proper string (for the bow).

Firearms : The chief distinction of the modern military science is the extensive employment of fire-arms, their invention being attributed to the Europeans, and it being supposed that fire-arms were unknown in ancient India. Nothing, however, is farther from the truth. Though the Indian masterpieces on the science of war are all lost, yet there is sufficient material available in the great epics and the *Purāṇas* to prove that firearms were not only known and used on all occasions by the Indians, but that this branch of their armoury had received extraordinary development. In mediaeval India, of course, guns and cannons were commonly used. In the twelfth century we find pieces of ordinance being taken to battle-fields in the armies of Prithviraj. In the 25th stanza of *Prithvirāja Raso* it is said, "The cavillers and cannons made a loud report when they were fired off, and the noise which issued from the ball was heard at a distance of ten *kosa.*"

नृप पंग नयर छूटे अराब।
कोटह कंगूर चढ़ि चढ़ि सिताब।।
जंबूर तोप छूटहि झंनकि।
दश कोश जाय गोला मनकि।।
सिरदार भार बाराह रोह।
लंगी अभंग बर हनै कोह।।

An Indian historian, Raja Kundan Lall, who lived in the court of the king of Oudh, says that there was a big gun named lichmā in the possession of His Majesty the King (of Oudh) which had been originally in the artillery of Mahārājā

Prithviraj of Ajmer. The author speaks of a regular science of war, of the postal department, and of public or common roads. See *Muntakhab Tafsee-ul-Akhbar*, pp. 149, 50.

"Maffei says that the Indians far excelled the Portuguese in their skill in the use of fire-arms."[32]

Another author quoted by Bohlen speaks of a certain Indian king being in' the habit of placing several pieces of brass ordnance in front of his army.[33]

Bullets : Bullets were made of iron with other substances inside or without any such substance. For smaller guns, the bullets were made of lead or any other metal.[34]

Bullets were fired by firing the gunpowder in the barrel.[35]

Guns : Here it may also be pointed out that guns, cannons and gun-powders were also in vogue in warfare in ancient India.

Guns were made of iron or some other metal. They were rubbed and cleaned daily and covered by armed men. According to Śukranīti (4.7.209-210), before use, the instrument has to be cleaned first, then gunpowder has to be put in, then it is to be placed lightlly at the origin of the instrument by means of the rod. Then the bullet has to be introduced afterwards gunpowder at the ear followed by application of fire to this powder, and the bullet is projected to the objective.

"Faria-e Souza speaks of a Gujrat vessel in AD. 1500 firing several guns at the Portuguese[36], and of the Indians at Calicut using fire vessels in 1502, and of the Zamorin's fleet

---

32 Hist. *Indica*, p. 25

33 *Das Aite Indien*, Vol. II, p. 63

34 सीसस्य लघुनालार्थे ह्यन्यधातुप्रभवोऽपि वा। शुक्रनीति- 4.7.204

35 कर्णचूर्णाग्निदानेन गोलं लक्ष्ये निपातयेत् । शुक्रनीति- 4.7.211

36 Asia Portuguesa, Tom I, Part I, Chapter 5.

carrying in the next year 380 guns."[37]

Cannons : In the description of Ayodhyā is mentioned the fact of *yantras*[38] being mounted on the walls of the fort, which shows that cannons or machines of some kind or other were used in those days to fortify and protect citadels.

The *Rāmāyaṇa*, while describing the fortifications, says "As a woman is richly decorated with ornaments, so are the towers with big destructive machines."[39] This shows that cannons or big instruments of war like cannons, which discharged destructive missiles at a great distance, were in use at that time.

Rockets : "Rockets," says Professor Wilson, "appear to be of Indian invention, and had long been used in native armies when Europeans came first in contact with them."

Col. Tod says, "Jud Bhan (the name of a grandson of Vajra, the grandson of Krishna), 'the rocket of the Yadus,' would imply a knowledge of gun-powder at a very remote period."

Rockets were unknown in Europe till last century (1906 AD.). "We are informed by the best authorities that rockets were first used in warfare at the siege of Copenhagen in 1807."[40] Mr. Elliot says, "It is strange that they (rockets) should now be regarded in Europe as the most recent invention of artillery."[41]

Machine throwing liquids

There were in ancient India machines which, besides throwing balls of iron and other solid missiles, also threw

---

37 *Ibid*, Chapter 7

38 *Yantra* means 'machine with which weapons were operated.'

39 *Rāmāyaṇa*, *Sundara Kāṇḍa*, Third Chapter, 18th verse.

40 Tod's *Rajasthan*, Vo. II, p. 220

41 *Penny Encyclopaedia*, V, 'Rocket.'

particular kind of destructive liquids at great distances. The ingredients of these liquids are unknown: their effects, however, are astonishing.

Ctesias[42], Elian[43] and Philostratus[44] all speak of an oil manufactured by Indians and used by them in warfare in destroying the walls and battlements of towns that no "battering rams or other polioretic machines can resist it," and that "it is inextinguishable and insatiable, burning both arms and fighting men."

Lassen says, "That the Hindus had something like 'Greek fire' is also rendered probable by Ctesias, who describes their employing a particular kind of inflammable oil for the purpose of setting hostile towns and forts on fire."[45]

Eusebe Salverte, in his Occult Sciences, says, "The fire which burns and crackles on the bosom of the waves denotes that the Greek fire was anciently known in Hindustan under the name of *baḍavā*."[46]

Gunpowder : It has also been noticed that gun-powders were also in vogue in warfare in ancient India alongwith cannons and guns. VD (67) specifically points out the existence *Rañjaka* (gun-powder). Accordingly, for the protection of throne of the king lot of gunpowder should also be stored. Gunpowder in ancient India was known by the name of *Rañjaka*. In *Jāmadagneya Dhanurveda* gun powder has been called as *Cūrṇa*. In *Śukranīti*, the same has been described as *Agnicūrṇa* or *Agni*.

In *Śukranīti*, the method and chemical composition for preparation of gunpowder has been given. Accordingly, five

---

42 Ctesie, *Indica Exerpta*, XXVII (ed. Baer), p. 356

43 *De Natura Animal*, *Lib*. V., cap. 3

44 *Philostrati Vita Apollonu*, Lib. III, cap. 1

45 Lassen's *Ind. Alt.* II, p. 641

46 English Translation, Vol. II, p. 223

palas (582.5 miligram) *suvarci* salt, one *pala* (11.5 miligram) of sulphur and one *pala* (11.5 miligram) each of the charcoal received from the wood of *arka* (swallow wart or asclepias gigantia), *snuhi* and aṅgāra plants by the Ayurvedic process of *Sadhūma Puṭapāka* where a drug is prepared in a closed vessel placed in a pit. The above mentioned salts and charcoals should be purified, powdered and mixed together. This mixture should then be soaked into the juices of *snuhi*, *arka* and garlic. It should be dried up in the sun and finally powdered like sugar, the substance will be gunpowder.[47]

In another type of preparation, the substances are the same, but the quantity of *survarci* salt is changed into six or four parts quantity of sulphur and charcoal remain the same.[48]

Experts also made gunpowder in various other ways of white and other hues of their choice according to the relative quantities of constituents like *Aṅgāra*, shulphur, *survarci* salt, *śilājit*, *haritāla*, left out lead after being purified, hiṅgula, ironfilings, camphor, *jatu*, indigo and juice of *sarala* tree etc.[49]

Professor Wilson says, "Amongst ordinary weapons one is named *vajra*, the thunderbolt, and the specification seems to denote the employment of some explosive projectile, which could not have been in use except by the agency of something like gunpowder in its properties."[50]

---

47 स्नुह्यर्काणां रसोतस्य शोषयेदातपेन च।
पिष्ट्वा शर्करवच्चैतदग्निचूर्णं भवेत् खलु।। शुक्रनीति- 4.7.202

48 सुवर्चिलवणाद्भागाः षड् वा चत्वार एव वा ।
नालास्त्रार्थाग्निचूर्णे तु गन्धाङ्गारौ तु पूर्ववत्।। शुक्रनीति- 4.7.203

49 अङ्गारस्यैव गन्धस्य सुवर्चिलवणस्य च
शिलाया हरितालस्य तथा सीसमलस्य च।
हिंगुलस्य तथा कान्तारजसः कर्पूरस्य च।
जतोर्नील्याश्च सरलनिर्यासस्य तथैव च। शुक्रनीति- 4.7.194

50 Wilson's Essays, Vol. II, p. 302. The Indians are from time immemorial remarkable for their skill in fireworks. The display of fire works has been from olden days a feature of the Dassehra

As regards "gunpowder," the learned Professor says, "The Indians, as we find from their medical writings, were perfectly well acquainted with the constituents of gunpowder sulphur, charcoal, saltpetre, and had them all at hand in great abundance. It is very unlikely that they should not have discovered their inflammability, either singly or in combination. To this inference *a priori* may be added that drawn from positive proofs, that the use of fire as a weapon of combat was a familiar idea, as it is constantly described in the heroic poems."[51]

The testimony of ancient Greek writers, who, being themselves ignorant of fire-arms used by Indians, give peculiar descriptions of the mode of Indian warfare is significant. ''The mistius mentions the Brāhmaṇa fighting at a distance with lightning and thunder."[52]

Alexander, in a letter to Aristotle, mentions "the terrific flashes or flame which he beheld showered on his army in India." See also Dante's Inferno, XIV, 31-7.

Speaking of the Indians who opposed Alexander the Great, Mr. Elphinstone says, "Their arms, with the exception of fire arms, were the same as at present."[53]

Philostratus thus speaks of Alexander's invasion of the Punjab "Had Alexander passed the Hyphasis he never could have made himself master of the fortified habitations of these sages. Should an enemy make war upon them, they drive him

---

festival. Mr. Elphinstone says, 'In the Dassehra ceremoney the combat ends in the destruction of Laṅkā amidst a blaze of fireworks which' would excite admiration in any part of the world. And the procession of the native prince on this occasion presents one of the most animating and gorgeous spectacles ever seen.' Elphinstone's *History of India*, p. 178.

[51] Essays, Vol. II, p. 303

[52] Orat, XXVII, p. 337. See Ap. Duten's *Origin of the discoveries attributed to the Moderns*, p. 196.

[53] Elphinstone's *History of India*, p. 241

off by means of tempests and thunders as if sent down from Heaven. The Egyptian Hercules and Bacchus made a joint attack on them, and by means of various military engines attempted to take that place. The sages remained unconcerned spectators until the assault was made, when it was repulsed by fiery whirlwinds and thunders which, being hurled from above, dealt destruction of the invaders."[54]

The western scholars thinks that the gunpowder was invented for the first time in Europe in 14th century AD. First of all the name of German monk Barthold Schwarz who dabbled in the black art of alchemy is quoted behind the invention of gun-powder. But later in the *Encyclopaedia Britannica* (P.5), it was stated that gun powder was either discovered in China or India. According to *Encyclopedia Columbia* (p.896), gun powder was invented in China. The knowledge of gun powder reached Europe through Arabs. But nobody tried to know how the Arabs received this knowledge. India is the source of all knowledge and science that reached Arab.

Hasta Mukta (Mannually operated weapons)

Following are hand-operated (manually operated) weapons used for close range of fighting.

1. Paṭṭiśa
2. Śakti, Vyjanti, etc.
3. Mugdar
4. Bhuśuṇḍī
5. Aśni
6. Daṇḍa
7. Akṣurapri̅
8. Ṛṣṭi
9. Vatsadanta
10. Asi
11. Tomar
12. Parigh
13. Musala
14. Kaṁpana
15. Karaṇī
16. Vārāhakarṇa
17. Bhindipāla
18. Silimukh
19. Vipāṭa
20. Aṅkuśa

---

54 Philostrati Vit: Apollon, *Lib* II, C. 33

More detailed description of some of mannually operated weapons is given below :

Daggers : Warlike weapons and splendid daggers were presented at the International Exhibitions of 1851 and 1862, and a critic speaking of them, says, "Beautiful as the jewelled arms of India are, it is still for the intrinsic merit of their steel that they are most highly prized."[55]

Swords : That the ancient Indians were celebrated for their sword fight is evident from the Persian phrase, "to give an Indian answer," meaning "a cut with an Indian sword." The Indian swordsmen were celebrated all over the world. In an Arabic poem of great celebrity, known as *Sabaa Moalaqa*, there occurs the passage, "The oppression of near relations is more severe than the wound caused by a Hindu swordsman."[56]

Ctesias mentions that the Indian swords were the best in the world.[57]

Explosives : In addition to the above weapons, ancient Indians also used used explosive weapons. Commenting on the stratagem adopted by King Hal in the battle against the king of Kashmir, in making a clay elephant which exploded, Mr. Elliot says, "Here we have not only the simple act of explosion but something very much like a fuze, to enable the explosion to occur at a particular period."[58]

Viśvāmitra, when giving different kinds of weapons to Rāma, speaks (in the *Rāmāyaṇa*) of one as *Āgneya*, another as *Śikhara*.

आग्नेयमस्त्रन्दयितं शिखरन्नाम नामतः।

---

[55] Manning's *Ancient and Mediaeval India*, Vol. II, p. 365

[56] The Tafsir Azizi says: Teghi-i-Hindi va Khanjar-i-roomi-Na Kunad aanki intiazar Kunad.

[57] Max Dunker's *History of Antiquity*, Vol. IV, p. 436

[58] Elliot's *Historians of India*, Vol.I, p. 365

"Carey and Marshman render *Śikhara* as a combustible weapon."[59]

In the *Mahābhārata* we read of "a flying ball emitting the sound of a thunder-cloud which is expressed in referring to artillery."[60]

The *Harivaṁśa* thus speaks of the fiery weapon:

आग्नेयमस्त्रं लब्ध्वा च भार्ग्गवात्सगरो नृपः।
जिगाय पृथिवीं हत्वा तालजङ्घान्सहैहयान्॥

"King Sagara having received fire-arms from Bhārgava conquered the world, after slaying the Tāljaṅghas and the Haihayas."

M. Langlois says that "these fire-arms appear to have belonged to the Bhargavas, the family of Bhṛgu."[61] Again,

उर्ध्वस्तु जातकर्म्मादि तस्य कृत्वा महात्मनः॥
अध्याप्य वेदानखिलांस्ततोऽस्त्रम्प्रत्यपादयत्।

आग्नेयन्तु महाबाहुरमरैरपि दुस्सहम्॥
स तेनास्त्र बलेनाजी बलेन च समन्वितः।

"Aurva having performed the usual ceremonies on the birth of the great-minded (prince), and having taught him the Vedas instructed him in the use of arms; the great-armed (Aurva presented him the fiery weapon, which even the immortals could not stand."

---

59 Various kinds of weapons are mentioned, some of which are extraordinary. As it is not known how they were made, what they were like, and how they were used, people think they are only poetic phantasies. Mr. Elliot says, ''Some of these weapons mentioned above were imaginary, as for instance, the vāyava or airy." But who would not have called the gramaphone, the cinematograph and wireless telegraphy imaginary only 50 years ago

60 Bohlen, *Das Alte Indien*, II, 66

61 *Harivaṁśa*, p. 68

Physics in Mahābharata

The World University Encyclopedia defines physics as the branch of science which sets as its object the ability to predict the behavior of natural phenomena with the help of a system of laws derived from observations and experiences. Physics is a quantitative science. There are two main branches of Physics – (A) Experimental and (B) Theoretical. Experimental physics is the science of observation and experiment which gives accurate knowledge of actutal behavior of natural systems. Theoretical physics builds up a system of quantitative relations among measured quantities and formulates these relations with the help of mathematics into physical laws.[62] Physics generally studies the area of matter, force, light, magnetism, motion, heat, electricity and sound. Hereunder, a brief introduction to the Vedic laws of physics will be given before we discuss physics known to the Mahābhārata scientists.

Vedic Law of Causation

According to the Vedic science, Brahman is the Nimitta-kāraṇa or efficient cause of creation, prakṛti the upādāna (samvāyī) or inherent cause/material cause of creation and soul, time and space are sādhāraṇa (asamvāyi-kāraṇa) non-inherent or simple cause of the creation. Here it may be known that the presence of efficient cause is essential for the creation of effect. In other words, without efficient cause no effect can be produced even when the material and simple causes are present. Here it may also be informed that the qualities of Nimitta-kāraṇa or efficient cause and sādhāraṇa (simple) cause are not inherited by the effect. Effect only inherits the qualities of upādāna (samvayi-karaṇa) or material cause. This whole phenomenon may be understood with the help of the following example. Suppose a potter makes a pot. He uses mud and potter's wheel in making the pot. In this condition, potter will be the efficient cause of pot, without whom production of pot

---

[62] World University Encyclopaedia. Vol. 9 Inc. Washington

is not possible. Mud will be material cause, the quality of mud will be inherited by the pot. Potter's wheel, time and space are the simple causes of the creation of pot. An efficient cause cannot have its cause, but a material cause of an effect may be the effect of some other material cause and this chain will continue till we reach atom. So in case of material or upādāna cause (samvāyi-karaṇa) atom is considered as having no cause to effect it. According to the Vaiśeṣika science handed down to us by the sage Kaṇāda, the matter is made of the smallest and indivisible tiny particle called paramāṇu. In the language of modern physics, it may be equated to single neutral particle called atom).

Vedic Principle of Nityatva

As per laws of Vedic physics, any dravya that has no material cause is considered as eternal. Matter in gross form is not eternal but in atomic form matter is also eternal. Eternal dravyas cannot be seen with naked eyes. God, soul mind, time and space are also eternal like atom of prakṛti (matter) or energy. Eternal dravyas can be perceived by a Yogīs or Siddha Puruṣas. A yogī is higher even than a siddha puruṣa. A Siddha Puruṣa is he who has equipped himself with a technology, spirtual or material, to perceive eternal matter like atoms, time and space. A yogī can have a perception of his own self and supreme self.

Eternal dravyas are known as sat and non-eternal dravyas are called asat. As per Vedic principle of nityatva, sat or nitya (eternal) padārthas can neither be created nor destroyed, but anitya or asat (non-eternal) dravyas can be created from their material cause and destroyed to merge with the material cause. The principle of 'nityatva' in Vedic Science is forerunner to the first law of thermodynamics of modern physics dealing with 'Conservation of Energy', which states that energy can neither be created nor destroyed in an isolated system. Thus atom can neither be destryed nor created, but it is the inherent or material cause of matter and matter of this creations of the visible universe. Since matter is the effect of atom. It can be created and destroyed. Prakṛti in Mahbhārata is called kṣara, that is convertible. In fact prakṛti converts into matter and matter into Prakṛti.

According to Vaiśeṣika science, a Dravya (substance) or Guṇa (quality) causes another substance or quality respectively of its own class.

द्रव्यगुणयोः सजातीयारम्भकत्वं साधर्म्यम् ।। 9 ।।

When a Dravya or Guṇa cause another Dravya or Guṇa of its own class, that is called Sādharmya (causing a substance of its own class).

Mahabhārata scientists also new that a dravya creates dravya; just as a seed produces a seed and body, a body.

द्रव्याद् द्रव्यस्य निर्वृत्तिरिन्द्रियादिन्द्रियं तथा।
देहाद् देहमवाप्नोति बीजाद् बीजं तथैव च ।। महा. शान्तिपर्व. 305.21

Vedic Principle of Disorder

Everthing in this visible world is vikṛti (evolute) of prakṛti (atomic matter/energy). Prakṛti means matter in original form (atomic form) and vikṛti means evolutes or matter in gross form. Saṅkya describes prakṛti as.

सत्त्वरजस्तमसां साम्यावस्था प्रकृति

sattva rajastamasāṁ sāmyāvasthā prakṛti

That is, prakṛti is homogenous state or state of equalibrium or orderly state of its properties-sattva, rajas and tamas. Vikṛti means disorderly state or heterogenous state or state of disequalibrium of sattva, rajas and tamas. Prakṛti (atom) has three properties–sattva (intelligence), rajas (motion) and tamas (inertia) which are known as electron, proton and neutron respectively. Order of sattva, rajas and tamas is called prakṛti. Disorder or disequalibrium of sattva rajas and tamas is called as vikṛti or devolution which is described in modern physics as evolution. At this state atoms start joining to form molecules and so on for the manifestation of visible world. According to Vedic physics, dissolution is the tendency of solid state of matter towards atomic state. This may also be described as tendency towards the orderly state sattva rajas and tamas. Vaiśeṣika Darśana says that reverse order starts with heat. Heat converts solid matters into liquids and so on and so forth. This reverse process in modern phsycis is called as disroder or decay.

Agni and Soma

Accordidng to the Vedic science, the whole world is the interplay of agni (-ve) or anti matter and soma (+ve) or matter.

अग्निसोमात्मकमिदं सर्वम् ।

In Mahābhārata, Maheśvara describes Umā 'Agni' (antimatter) and 'Soma' (matter) as two types of nature of the constituents of creation of visible world.

तदहं कथयिष्यामि शृणु तत्वं समाहिता।
द्विविधो लौकिको भावः शीतमुष्णमिति प्रिये।। महा. अनु.पर्व 141

Force & Action

Force is an external agent capable of changing the state of rest or motion of a particular body.

Mahabhārata scientists were informed of the fact that force can accelerate the motion.

बलप्रयत्नादधिरूढवेगाम् । महा. शल्यपर्व. 17.47

According to them, force and action are interdependant. A force without action is useless and action without force cannot work.

यथा बलं क्रियाहीनं क्रिया वा बलवर्जिता।
नेह साधयते कार्यं समायुक्ता तु सिध्यति।। महा. अनु.पर्व 108.20

Gravitional force of earth

Gravitational force of earth was recognised in the Vedic period itself. It was known by many names, such as ākarṣaṇa, ākṛṣṭi śakti or gurutva or gurutva śakti. During the period of Mahābharata (Śānti parva, 255.3) also gravitational force of earth was well known. Bhīṣma Pitāmaha while counting the 10 properties of the earth before Yudhiṣṭhira also enumerates gurutva śakti (gravitational force) as one of the property of the earth.

भूमेः स्थैर्यं गुरुत्वं च काठिन्यं प्रसवार्थता।
गन्धो गुरुत्वं शक्तिश्च संघातः स्थापना धृतिः।। महा. शान्तिपर्व 255.3

In fact, the word 'gravity' is formed of Sanskrit word 'gurutva'. It shows that ancient Indians were not only aware of gravitational force of the earth, but they made it known to the other parts of the world.

Famous mathematician and astronomer Bhāskarācharya (2nd century B.C) in his celebrated book Siddhānta Śiromaṇī (Golādhyāya, Bhuvanakośa, 5-6) describes the gravitional pull of the earth as under:

आकृष्टिशक्तिश्च मही यथायत् खस्थं गुरु स्वाभिमुखं स्वशक्त्या।
आकृष्यते तत् पततीव भाति समे समन्तात क्व पतत्वियं खे।।

[Meaning] The earth has gravitational pull due to which it attracts the things to it. Due to gravitational pull, the things fall upon it.

Mirror

In the Mahābhārata, we find the frequent use of the word 'ādarśa' and 'darpaṇa' which makes it evident that mirror was known to ancient Indians since remote period. Some references are appended below:

प्रतिबिम्बमिवादर्शे गुरुपत्न्याः शरीरगम्।
स तं धोरेण तपसा युक्तं दृष्टा पुरन्दरः।।
प्रावेपत सुरांत्रस्तः शापभीतस्तदा विभो। महा. अनु.पर्व. 41.18

भार्यायां जनितं पुत्रमादर्शेष्विव चाननम्।
ह्लादते जनिता प्रेक्ष्य स्वर्गं प्राप्येव पुण्यकृत्।। महा. आदिपर्व 74.49

यथा हि पुरुषः पश्येदादर्शे मुखमत्मनः।
एवं सुदर्शनद्वीपो दृश्यते चन्द्रमण्डले।। महा.भीष्म.पर्व 5.16

Convex lens

During Mahābhārata period (3101 B.C.) Convex lens was known as 'Sūryakānta maṇi'. The people during that period were well aware that a convex lens can produce fire using sunlight, as a convex lens focus light to one point. Convex lense concentrates the light energy to one spot on a thing so that the heat energy accumulates on that one small spot and burn things.

यथाऽऽदित्यान्मणेश्चापि वीरूद्भयैँव पावकः।
जायन्त्येवं समुदयात् कलानामिव जन्तवः।। महा. शान्तिपर्व 320.126

रेतो वटकणीकायां घृतपाकाधिवासनम्।
जातिः स्मृतिरयस्कान्तः सूर्यकान्तोऽम्बुभक्षणम्।। महा.शान्तिपर्व 218.29

Ancient Vedic scholar Yāska (2900 BC) in his famous treatise Nirukta (7.6) also refers to production of fire by convex lens. According to him, when sun is perpendicular, one can burn the dried cow dung cakes by focusing sunlight through maṇi (convex lens) on it.

अथादित्यात्–प्रदीचिप्रथम समावृत्ते आदि कंसं वा मणि वा परिमृज्य प्रतिस्वरेयत्र गोभयमसंस्पर्शयन् धारयति, तत् प्रदीप्तते, सोऽयमेव सम्पचते।

Magnetism

The history of magnetism dates back earlier than 600 BC, but it is only in the twentieth century that scientists have begun to understand it, and develop technologies based on its understanding. It is conjectured that magnetism was the most probably first observed in a form of the mineral magnetite called lodestone, which consists of iron oxide-a chemical compound of iron and oxygen. The ancient Greeks were the first known to have used this mineral, which they called a magnet because of its ability to attract other pieces of the same material and iron.

The Englishman William Gilbert (1540-1603) was the first to investigate the phenomenon of magnetism systematically using scientific methods

But magnet and magnetism was known in anceint India before the Greeks came to know about it. We find a mention of magnet in the name of 'Ayaskānta Maṇi' and its properties 5000 years ago in the Mahābhārata, Śānti parva (211.3). There it is mentioned that as a lifeless piece of iron is attracted to the magnet, similarly, when the body is born in a particular species, a concerned creature is automatically attracted to the natural instincts, temperament and qualities attributing that species.

अभिद्रवत्ययस्कान्तमयो निश्चेतन यथा।
स्वभावहेतुजा भावा यद्वदन्यदपीदृशम्।। महा. शान्तिपर्व 211.3

Vaiśeṣika Darśana (5.1.25) which dates before Mahābhārata period (3101 BC) also makes the mention of compass needle and says that the motion of compass needle is due to invisible cause, which cannot be defined as per any rules of physics.

मणिगमनं सूच्यभिसर्पणमदृष्टकारणकम्।

According to Rasārṇava (12th century AD), there are five basic varieties of magnet. They are bhrāmaka, chumbaka, karṣaka, drāvaka and romakānta. These five varieties are further devided into 6 sub-varities each like single faced, two faced, three faced, four faced, five faced and multi-faced. All these thrity varieties are avalable in 3 colours-yellow, red and black. Thus by the time of Rasārṇava, 90 varietes of magnet were known and used by Indians.

भ्रामकं चुम्बकं चैव कर्षकं द्रावकं तथा।
एवं चतुर्विधं कान्तं रोमकान्तं च पंचमम्।।
एकद्वित्रिचतुः पंचसर्वतोमुखमेतत्।
पीतं कृष्णं तथा रक्तं स्यात् पृथक् पृथक्।। रसार्णव, 6.40.41

From the foregoing discussion, it is evident that magnet and magnetism is the discovery of Indians much before the people in other parts of globe came to know about it.

Heat transfer

Mahābhārata scientists were aware of the principle that heat is normally transferred from a high temperature object to a lower temperature object. Heat transfer changes the internal energy of both systems involved. Quoting the statement of King Janak before Yudhiṣṭhira, Śaunaka says, when mind is disturbed, body is also grieved, like the hot ball of iron if put in the water pot heats up the water too.

मानसेन हि दुःखेन शरीरमुपतप्यते।
अयः पिण्डेन तप्तेन कुम्भसंस्थमिवोदकम्। महा. वन पर्व 2.25

Electricity

Electricity is considered as the greatest invention in history because it opened people up to a whole new world. Without power, the world would never be able to innovate. Also as time goes on people continue to expand this invention and innovate it. Most inventions would have never happened without electricity. Since it was invented, most inventions were based on it and it was used to help create the invention. Electricy is is not the invention of modern world, but ancient Indians were the first inventor of it. In the Vedas, electricity has been mentioned by the name Indra. The word *Vidyuta* is also used for the same. We also find the mention in *Agastya Saṁhitā* regarding the method of generation of electricity and constructing electrical batteries. The relevant reference can be quoted as under:

संस्थाप्य मृण्मयेपात्रे ताम्रपात्रं सुसंस्कृतम्।
छादयेच्छिखिग्रीवेण चाद्राभिः काष्ठपांशुभिः।।
दस्तालेष्टो विधात् पारदाच्छादितस्ततः।
संयोगाज्जायते तेजो मित्रावरुणसंज्ञितम्।।
अनेनजलभंगोऽस्ति प्राणोदानेषु वायुषु।
एवं शतानां कुम्भानां संयोगकार्यकृत्स्मृतः।।

sansthāpya mṛṇmaye pātre tāmra-patraṁ susanskṛtam

chādayec chikhigrīvena ca ārdrābhiḥ kāṣṭhapānsubhiḥ

dastāleṣṭo vidhāt pārad ācchāditas tataḥ

sañyogāt jāyate tejo mitrā-varuṇa-sañjitam

anena jalabhaṅgo'sti pāṇa udāneṣu vāyuṣu

evaṁ śatānāṁ kuṁbhānāṁ sañyoga kārya kṛt smṛtaḥ.

"Place a well-cleaned copper plate in an earthenware vessel. Cover it first by copper sulfate (*śhikhigriva*) and then moist sawdust (*kāṣṭhapānsu).* After that put a mercury-amalgamated

-zinc (*pārad ācchādita dast)* sheet on top of an energy known by the twin name of Mitra-Varuna (cathode and anode). Water will be split by this current into Praṇa-vāyu (oxygen) and Udāna-vayu (hydrogen). A chain of one hundred jars is said to give a very active and effective force to make a battery bank (**(kārya kṛt)**)."

Further it is stated that

वायुबन्धकवस्त्रेण निबद्धो यानमस्तके।
उदानः स्वलघुत्वेन बिभर्ति–आकाश–यानकम्।

vāyu-bandhak-vastreṇa nibaddho yāna-mastake.

uādānaḥ svalaghutvena bibharti-ākāśa-yānakam.

"That is to say this energy can be used in extracting hydrogen (udāna-vāyu) from water and then the same, being lighter than air can be used in the operation of an aeroplane."

Apart from this, the *Atharvaveda* (20.7.2) refers to two arms of electircity. These two arms are nothing else but two charges of electricity i.e. positive and negative along with its friendly and destructive use. The mantras goes like this.

नव यो नवतिं पुरो बिभेद बाह्वोऽजसा। अहिं च वृत्रहावधीत्।।

nav yo navatiṁ puro bibheda bahvo'jasā.

ahiṁ ca vṛtrahāvadht.

Electricity, which breaks, by the energy of its arms (i.e. positive and negative currents) the 99 cities called bhogas in the Vedic terminology (so-called elements known to modern scientists), destroys the cloud, which makes clouds precipitate, the source of energy and power.

In the Mahābhārata, we come across an Ākhyāna related to Aṣṭāvakra which refers to a debate between King Janaka and Aṣṭāvakra. In this debate Janaka pose a question to Aṣṭāvakra with reference to atmospheric electricty as, who (electricity) remains united like two mares (+ve & -ve charges) and falls from sky like falcon? Which of the devas conceive them and both of them make whom to deliver?

वऽवे इव संयुक्ते श्येनपाते दिवौकसाम्।

कस्तयोर्गर्भमाधत्ते गर्भं सुषुवतुश्च कम्।। महा. वनपर्व 133.26

Aṣṭāvakra gives reply: O Rajan! let them not fall upon even on the residence of your enemies. They are driven by wind and clouds conceive them in their womb and both of them make clouds to deliver (rain).

मा स्म ते ते गृहे राजंच्छात्रवाणामपि ध्रुवम्।
वातसारथिरागन्ता गर्भं सुषुवतुश्च तम्।।महा. वनपर्व 133.27

In the above description of atmospheric electricity two mares signify -ve and +ve charges. Not only this, the above description also shows that ancient Indians were very well familiar with the process of rainformation from clouds, i.e. the rain is formed by the neutralisation reaction process between the -vely charged clouds and +vely charged clouds or -vly charged clouds and neutral earth. For deatils on 'Vedic science of Rainmaking', see 'Vedic Meteorology' a book written by the present author.

According to the Ṛgveda (1.16.5), the electricity is hidden in the water and when it comes out, it spreads light and provides energy. *Ṛgveda* (1.85.5; 188.1) describes the use of electricity in weapons and telegraphy. The word *Vidyut* is used to denote electricity in Vedas. Electricity can be generated from energy (*RV.* 1.45.5). Electricity protects people and should be used as destructive energy against wicked persons and enemies with the help of weapons, which work on electricity (*RV.* 1.86.9). A scientist who knows the nature of Time and all the properties and characteristics of electricity can accomplish his/her work very fast (*RV.* 1.95.8).

The Atharvaveda[63] covers electricity as one of the most exciting topics. It discusses a detailed description of valuable applications that harness and utilize this immense source of energy. Some of the applications described include important specifications for a control system that harnesses the intense

---

[63] See for detail, Engineering and Technolgy in Ancient India, Ravi prakash Arya, Indian Foundation for Vedic Science, 2020. (P:154)

power of electricity for use as a deadly weapon, utilization of hydroelectric power for manufacturing, and the fission properties of electricity.

Vedic Scientist Rao Sahab Vajhe[64] has classified electricity into 6 types based upon the ancient treatises like Agastya Saṁhitā. They are as under:

1. Taḍita: produced by rubbing two silk cloths
2. Saudāminī: Produced by rubbing two jewels
3. Vidyut: Atmospheric electricty
4. Śatakuṁbhī: Electric current produced by cells
5. Hradani: Stored electricity
6. Aśani: Electric current created by magnet

On the basis of foregoing discussion, it can be inferred that ancient Indians studied deeply all kinds of electricity and applied it for their benefit.

Electric bulb

From the references in the Mahābhārata, we are given to understand that electric bulb or lamp was in use in ancient India. It was called 'Nivāta dīpa' i.e. 'lamp or bulb burning on non-reactive gas'. The references are cited below:

यथा दीपो निवातस्थो निरिंगो ज्वलते पुनः।
तथासि भगवन् देव पाषाण इव निश्चलः।। महा. शान्तिपर्व 46.6

Just as lamp bruning on non-reactive gases does not flicker, similarly, O Bhagavan, you are stable like rock.

लक्षणं तु प्रसादस्य यथा स्वप्ने सुखं स्वपेत्।
निवाते वा यथा दीपो दीप्यमानो न कम्पते।। महा. शान्ति पर्व 246.11

Just as lamp bruning on non-reactive gases does not flicker, similarly non-fluctuation is the sign of stabilized mind.

This 'Nivāta lamp' was nothing but modern day bulb

---

[64] P.V. Hole, Machines in Sanskrit Literature.

burning on non-reactive or inert gases.

In Mahābhārata, we come across a reference of a city lighted by 6000 lamps buring on 'hrad' (electricity stored in cells). I quote the refrence below:

जलेश्वरस्तु ह्रत्वा तामनयत् स्वं पुर ˙ प्रति।
परमाद्भुतसंकाशं षट्सहस्रशतह्रदम्।। महा. अनु.पर्व 154.14

Varuṇa brought her to his city lighted by 6000 lamps or bulbs burning on electricity stored in cells.

www.ingramcontent.com/pod-product-compliance
Lightning Source LLC
Chambersburg PA
CBHW030019190526
45157CB00016B/3132